Constructive Adpositional Grammars

Constructive Adpositional Grammars: Foundations of Constructive Linguistics

By

Federico Gobbo and Marco Benini

CAMBRIDGE SCHOLARS PUBLISHING

Constructive Adpositional Grammars:
Foundations of Constructive Linguistics,
by Federico Gobbo and Marco Benini

This book first published 2011

Cambridge Scholars Publishing

12 Back Chapman Street, Newcastle upon Tyne, NE6 2XX, UK

British Library Cataloguing in Publication Data
A catalogue record for this book is available from the British Library

ISBN (10): 1-4438-3007-0, ISBN (13): 978-1-4438-3007-2

Table of Contents

List of Figures

List of Tables

Legenda

Abbreviations

adtree .... shortcut for ‘adpositional tree’

adj .... generic adjectitve

adp .... generic adposition

dep .... dependent group

gc .... generic grammar character

gov .... governor group

?the Lady ..... example of doubtful expression (syntactically or semantically)

*_good here_ .... example of ungrammatical expression (syntactically)

~~with~~ .... example of expression canceled by a transformation

Abs .... absolutive case (morphosyntactically marked)

Acc .... accusative case (morphosyntactically marked)

Gen .... genitive case (morphosyntactically marked)

Dat .... dative case (morphosyntactically marked)

Erg .... ergative case (morphosyntactically marked)

Nom .... nominative case (morphosyntactically marked)

Opener .... semantic role, case or concept in a semantic frame (example)

Symbols

← .... indicator of dependency

→ .... indicator of governement

↔ .... underspecified indication governement-dependency

> .... shortcut for transference

⇝ .... adpositional tree transformation

△ .... shortcut for a hidden adpositional tree

Letters

$\downarrow \mathcal{A}$ .... assertive pragmatic character
A .... adjunctive grammar character
$\uparrow \mathcal{C}$ .... commissive pragmatic character
D .... generic grammar character of the dependent
$\updownarrow \mathbb{D}$ .... Declaration (pragmatic character)
$\uparrow \mathcal{D}$ .... directive pragmatic character
E............. circumstantial grammar character
$= \mathcal{E}$ .... expressive pragmatic character
ϵ .... zero-marked adposition
F............. adposition grammar character imposed by the adposition
G .... generic grammar character of the governor
I .... underspecified or generic verbant grammar character
I_x^v .... verbant grammar character (v indicates valency, x saturation)
$\overleftarrow{\mathrm{I}}$ .... unaccusative verbant (always monovalent)
$\overrightarrow{\mathrm{I}}$ .... unergative verbant (always monovalent)
I^2 .... bivalent verbant grammar character
I^3 .... trivalent verbant grammar character
I^4 .... tetravalent verbant grammar character
I^5 .... pentavalent verbant grammar character
L............. listener (type of actant)
λ............. generic pragmatic character
O .... stative grammar character (extra-valency or generic)
O_x .... stative grammar character (x indicates the actant value)
O_1 .... stative grammar character (first valency)
O_2 .... stative grammar character (second valency)
O_3 .... stative grammar character (third valency)
O_4 .... stative grammar character (fourth valency)
O_x .... stative grammar character (extra valency)
Q .... extra in-valent actant in the construction (as O_x)
R .... receiver (a type of actant)
S............. speaker (a type of actant)
W............ fourth in-valent actant in the construction (as O_4)
X............ first in-valent actant in the construction (as O_1)
$\overleftarrow{\mathrm{X}}$............ unaccusative actant (always for $\overleftarrow{\mathrm{I}}$)
$\overrightarrow{\mathrm{X}}$............ unergative actant (always for $\overrightarrow{\mathrm{I}}$)
Y .... second in-valent actant in the construction (as O_2)
$\overleftarrow{\mathrm{Y}}$............ non-prominent second in-valent actant (in government)
$\overrightarrow{\mathrm{Y}}$............ prominent second in-valent actant (in dependency)
Z............. third in-valent actant in the construction (as O_3)

Chapter One

Introduction

This book presents a framework to understand how language is structured—in particular at a morphological and syntactic level—and how language takes its place in the general human cognition. [1]

We consider that the study of languages should be intertwined to the research in the field of cognitive sciences on one hand and with the expressive power of constructive mathematics on the other hand. In fact, constructive mathematics allows a full-blooded computational development of natural language grammar description, without invoking idealistic interpretations. In other words, every concept of our framework poses on a solid mathematical ground when interpreted in a cognitive way.

A preliminary clarification is needed. As the main area of investigation within this book is natural language grammar, we will use the term ‘language’ to refer to natural languages—such as English, Chinese, or Urdu—in general terms if the context is clear enough to avoid ambiguity. Otherwise, we will specify if we are talking about natural or formal (artificial) languages. Analogously, this kind of convention will be followed for other common terms which are used both by linguists and mathematicians in very different ways, such as ‘grammar’, ‘syntax’ or ‘semantics’. [2]

This clarification is needed because this book is addressed both to people belonging to the humanities—such as most linguists are, but also pedagogists involved in language learning issues—and people belonging to hard sciences, above all, mathematicians, but also logicians, computer scientists and natural language engineers. We believe that the so-called ‘two cultures’—even three, as recently posed by Kagan (2009)—should at least dialogue one with the other if not put in confluence for mutual improvement.

This is the main reason why we decided to put together concepts and results of our investigation, as they are at the same time linguistic and formal in nature. In other words, there is no dedicated chapter to the formal modelling [3] or to linguistic description, simply because the insights we have on one side have an immediate and direct effect to the other side, and writing in such a separate way would lead to unnecessary intricacies in order to let the reader understand our line of reasoning.

Of course, we do not pretend to reinvent the wheel; our work is well rooted both in the linguistic and mathematical traditions of their own. The sequel of this chapter is devoted to clarify what are our start points and how they are related (or not) with other works in the fields.

1.1 The adpositional paradigm

It is worth noticing that the term 'constructive' is used in different and complementary ways within the book. In this section, we explain 'constructive' referring to constructive mathematics. The fundamental concept of this book is the *adposition*, which is linguistically a generalisation of conjunctions, prepositions, postpositions, and so on—broadly speaking, linking words. The adpositional paradigm was the main result of the PhD dissertation by Gobbo (2009).

Since then, we put our work a step forward, and this book is the presentation of the results of that effort. In fact, now adpositions—the cornerstones of the adpositional paradigm—are described in the spirit of constructive mathematics, and that's why the framework is called 'constructive adpositional grammars' (Constructive AdGrams for short).

Constructive mathematics (Bridges and Richman, 1987) is, synthetically, a way to develop the mathematical thought that strictly preserves the information content of any statement. Precisely, disjunctive and existential statements are required to indicate witnesses for their truth: for example, $\exists x.P(x)$ can be proved in some theory T if and only if we can exhibit some value v such that $P(v)$ holds, i.e., it logically descends from T. So, some forms of logical thinking are not accepted, e.g., the Law of Excluded Middle, since they introduce unjustified information: in fact, if $P \vee \neg P$ is an axiom, it does not indicate whether the first or the second disjunct holds and there is no guarantee that such an information can be recovered from the theory T, and, in general, it cannot (Troelstra and Schwichtenberg, 2000; Troelstra, 1977).

The term 'grammars' refers to the fact that the mathematical structure is one, while the instantiations are many as natural languages are. We look at different shadows (i.e., natural languages), each one with its different shape, while the sun is always the same (the structure)—see Chapters 2 and 3 for details.

Before proceeding, it is important to explain how we deal with the 'shadows', i.e., the diversity of natural languages—in other words, the problem of linguistic universals and typology. From a phylogenic point of view, while every species seems to understand their members, only humans developed different languages—actually, more than 6,000. Furthermore, from an ontogenetic point of view, every child builds his own linguistic repertoire according to the environment he is grown up into, not his ethnic origin, and there are some regularities in the developing of language skills, from baby utterances to adult-like representations. According to Tomasello (2003), the most rigorous and plausible theory of language that characterises adult linguistic competence in child-friendly terms is rooted both in cognitive and social skills. From the cognitive side, the theory of mind has shown us that our ability to communicate is based on (a) intention-reading, i.e., the ability of inferring what the listener is expecting from us, beyond the literal meaning of what the speaker said; (b) pattern-finding, i.e., the ability to categorise our sensibilia in a mapping into the mind. From the social side, we learn the intentional actions of other humans by imitation.

These universals of learning (intention-reading, pattern-finding and imitation) applied to our innate linguistic ability are the fundamental elements of linguistic *constructions*—synthetically, patterns of usage of form-meaning correspondences that carry the messages beyond the gist of the words that took part in the construction itself. This fact leads us to another use of the term 'constructive' within this book.

Linguistic communication is symbolic, and symbols are "social conventions by means of which one individual attempts to share attention or mental state to something in the outside world" (Tomasello, 2003, 8). Symbols are used in patterned ways, and these patterns give form to linguistic *constructions*. Hence, constructions arise by two different forces: by the meaning of their parts, and by the frequency of use of the pattern itself. This species-unique process occurs over time and is called *grammaticalization*.[4]

The results of this cultural and historical process are natural language grammars, and that is why we have more than one. Therefore, grammars should be investigated as a result of the process of grammaticalization, not as static, monolithic entities. Most puzzling linguistic phenomena can be enlightened in terms of the grammaticalization process, as we will explain throughout this book. In fact, the frequency of pattern-usage leads to loose (hide) unnecessary information: that's why highly frequent words are usually short in terms of syllables.

Each grammar is made of constructions, providing us the linguistic data to be analysed. Constructions being central, our work can be put into relation with the early stages of transformational grammar (Chomsky, 1957, 1965) and to the works of cognitive linguists proposing constructional analyses, such as Tomasello (2003), Goldberg (1995, 2006), and Croft (2001).

Most of our linguistic observations derive from the works by cognitive linguists. The main limit of these analyses lies in the fact that they are *constructional* instead of constructive. That is, they take constructions as primitives, rejecting *in toto* any formalism, which is seen as inevitably 'Chomskyan' in nature. Matthews (1993) already observed:

> [...] for more of the past fifteen years, despite occasional disparagement from one side or another, each school has in practice had little reason to refer to the other. It is worth noting, for example, that Croft's recent introduction to *Typology and Universals* (1990) cites no work by Chomsky. (Matthews, 1993, 45)

The situation is even getting worse, as exemplified by the most radical constructional approach, which states that there is no linguistic category that is both formal and universal, as asserted by (Croft, 2001, 4):

> Of course, abandoning universal categories and relations leads to a very different view of Universal Grammar. Under the alternative view, Universal Grammar does not consist of an inventory of universal categories and relations available to all speakers. [...] The formal structures in grammars are language-particular, and universals of language must be sought elsewhere. [...] In principle, that appears to be the direction that Chomskyan generative grammar has headed: general constraints on syntactic structure but a proliferation of syntactic categories. In practice, however, the syntactic categories are assumed to be cross-linguistically valid, and the same categories (or a subset thereof) are posited of every language. This practice also holds for other formal syntactic theories.

We claim that a formalism is possible without neglecting the usage-based results of cognitive linguists. The problem is in the general perspective. Most cognitive linguists are 'maximalists', i.e., they consider semantics and syntax indissociable, and therefore semantics is put at the centre of analysis, while syntax is put at the periphery—traditionally, most results in cognitive linguistics are devoted to the semantic level, in particular the analysis of metaphors (Lakoff, 1997, for example). Conversely, in whatever (linguistic) formalism, the algebraic rules are insensitive to the meanings of the element they algorithmically combine, and hence linguistic meaning is put at the periphery of the system, the (morphosyntactic) rules being the core.

Nonetheless, we take into account the critique of later Chomskyan development (Chomsky, 1981, 1992) that constructions cannot be disregarded as epiphenomena, while the earlier Chomskyan models showed that the formalisation of constructions is not only possible but also feasible (Goldberg, 1995, 1). After all, Chomsky is a leading figure in the Cognitive Revolution, along with George Miller, Marvin Minsky, Allen Newell and Herbert Simon [5]: we claim that at least part of Chomsky's work can be put in the stream of cognitive linguistics, if we do not reject formalisation as a whole. At the same time, we believe that another way—not based on constituents—to formalise grammars is possible and worth attention, particularly for cognitive linguistics. [6]

In sum, it's no more the time of the "linguistic wars" (Harris, 1995): we claim that both approaches are valid, as they talk about different aspects of the same phenomenon, i.e., natural language grammars: the Chomskyan approach is top-down, deductive, because it looks for regularities beyond the variety of languages, while cognitive linguists try to explain variety on a usage-based perspective, following a data-driven, inductive approach. Our aim is to take the best practices from both approaches, without adhering to any 'Church'. Here we offer a strong, general, explicative formalism with a lot of examples taken from different languages of the world, with a special regard to English.

Our formalism takes into account the results of linguistic typology. Are there any language universals? The cross-analyses of grammars made by typologists showed us that grammar categories, arisen in Greek and Roman context for educational purposes—i.e., teaching and learning of Greek and Latin as written languages—cannot be forced as such into native languages of most part of Africa, Southeast Asia, the Americas and Australia. Should we look through the Procrustean lens of Standard Average European to those languages? Certainly not. Tomasello (2003) takes a radical conclusion, claiming that no universals of form exist, in particular no linguistic symbols, no syntactic constructions and no grammatical categories. If linguistic symbols highly depend on the socio-cultural context within they emerge (but what about the word *mama*?) and typological studies have shown that the variety of syntactic constructions is really impressive, nevertheless this variety *should* have a common, cognitive basis—and Greenberg's results in syntactic typology cannot be disregarded so easily. Tomasello claims that the only language universals that exist are the ones of communication and cognition, being in particular the presence of expressions of reference and the presence of predication. Moreover, they have no counterpart in terms of linguistic universals of form. We claim exactly the opposite: the mechanisms underlying the constructions are the linguistic counterparts of the cognitive ability of pattern-finding and intention-reading, which let the speaker and the listener—in the simpler, default case—to build up the linguistic representation within their joint attentional frame. The best account of the attention focus as a cognitive linguistic ability we have found in literature is the trajectory/landmark asymmetry by Langacker (1987, 1990, 1991), of which we will give a formal interpretation in terms of information prominence—see Chapter 2.

Attention should be directed somewhere in order to function properly: we claim that four grammar characters are universal, being the expressions of reference, predication, and their respective modifiers—respectively, adjuncts and circumstantials. Whorf (1945) and Tesnière (1959) did come to the same conclusion, even if starting from different bases. In particular, the work by Tesnière (1959) was the start of the so-called Dependency Grammars, which was rarely put into relation with cognitive linguistics.[7] We deeply analysed that classic work through the lens of modern formalisms and we have found that most, if not all, derived works have taken the concepts of dependency and verbal valency from Tèsniere, but, at the same time, completely disregarding his use of grammar characters.[8]

Moreover, we retain the concept of 'prepositional system', borrowed from Pennacchietti's analysis of Brøndal (1940): each natural language grammar has a relatively small set of prepositions (or other kind of adpositions) whose function is determined by the result of the opposition with the other prepositions in the grammar itself. In fact, even genetically close languages—such as French and Italian—show a considerable difference in the use of their prepositions because they belong to different prepositional systems (Pennacchietti, 2009, 2006).

The central role given to adpositions—broadly, prepositions, postpositions or in-positions, depending on the specific language—lead to another fundamental grammar character: adpositions. This grammar character collects the 'structural morphemes' of a language. Because of its somewhat technical nature, it has been neglected for a long time. The name 'adpositional paradigm' followed naturally.

The adpositional paradigm was first developed in Gobbo (2009), which extends Pennacchietti's work in an original way. Adpositions become a more abstract, general element in order to understand grammar structures. In particular, zero-marked adpositions—signed through an epsilon (ϵ)—put word order phenomena in the same realm of morphology. Thus morphology and syntax are clarified through a unified model and a unique mechanism—with some special features of their own. Semantics and pragmatics also have their place in the model, while phonetics and phonology have not.[9]

A remark made by an anonymous reviewer on the linguistic foundations of the adpositional paradigm was that the model was tested only into nominative-accusative grammar cases. We have taken this remark very seriously, and the present book fills this gap, providing a new constructive adpositional grammar model which takes into account both nominative-accusative and ergative-absolutive grammars. Our reference for ergativity under a theoretical point of view is the work by Dixon (1994), who notes in the appendix that, surprisingly, there are few theoretical models which gives an explanation of ergativity, because the theory "would have to recognise that there are three basic syntactic-semantic primitives (A, S and O) rather than just two ('subject' and 'object') however these are defined", see (Dixon, 1994, 236).

Constructive adpositional grammars solve this problem in terms of information prominence, as explained in the next chapters. Furthermore, we adhere to the unaccusative hypothesis, originally proposed by Perlmutter (1978), which gives a clear and convincing linguistic account of the problem, even if the Relational Grammar (pseudo)formalism is not convincing. [10]

Finally, no agreement is possible on a finite, fixed list of the universal types of semantic roles—called in a Chomskyan perspective 'theta-roles'—simply because they depend on the constructions they belong to. In other words, semantic roles are construction-dependent and hence language-defined, and so their list is open and undefined, at least for the purposes of the present book. [11]

For example, if a construction deals with food, such as *X eat Y* it has much more sense to call *X* Eater instead of Agent, following the tenets of cognitive linguistics, as the semantic roles belonging to the semantic frame of food activated by a constructions such as *X eat Y* show distinctive characteristics in the distributional analysis of the corpus in different languages—e.g., English, German, and Bengali (Croft, 2009)— which cannot be explained with a suitable degree of precision by the more used dichotomy Agent *vs.* Patient.

For example, the English construction *Y-eating X*, exemplified by *oil-eating microbes*, states that *Y* is the Food and *X* is the Non-human eater, prototypically having the sememe animal, so *oil-eating robot* is also acceptable, while *fish-eating child* is not, since *child* is not a Non-human eater. [12]

What it is important to underline here is that the *productivity* of constructions—i.e, how much *X*s and *Y*s can vary within the scope of that particular construction—is bound to the construction itself.

The aim of semantic roles is exactly to represent this degree of freedom, which can be zero at the limit, as in the case of completely grammaticalized idiomatic expressions—such as *kick the bucket*, where each lexical item cannot be moved paradigmatically—or very high, such as *Let's X!*, where the only constraint applied to *X* is its grammar character, in particular the need to belong to the class of English verbs.

1.2 The formal model

In the fields of mathematical and computational linguistics there are many natural language grammar formalisms currently under investigation. In particular, our formalism can be put into the realm of the so-called 'categorial grammars'—i.e., representations of natural language grammars in terms of categories. This line of research is far from being new, being rooted in the works by Ajdukiewicz (1935), Church (1940), Bar-Hillel (1953), and Lambek (1958). [13]

Our formal model is intended as a guiding reference for the development of linguistic concepts. In this sense, it should be understood as a 'weak' model which provides enough insight to understand the complexities of natural languages but not yet enough power to treat them in a computational fashion. The full details of the formal model are presented in mathematical terms in Appendix B: the reader is referred to that part for the formal details.

When the formal model is applied to any natural language, a parser, i.e., a computer program which allows to recognise the expressions belonging to the grammar, can be naturally derived, due to the constructive nature of the formal representation. [14]

The starting point is to consider any linguistic expression as a formal object, represented as an adpositional tree (adtree, for short) on morphemes. The precise structure of such trees is explained in the following chapters; for the moment being, we just need to know that each of them represent a unique piece of text in some natural language. Moreover, a tree represents a 'unique interpretation' of the corresponding text, i.e., a unique way to understand the grammar characters of each element in the text (the concept of 'grammar character' will be discussed and explained in detail in Chapter 3). For example, let's take the Italian sentence:

(1-1.) La vecchia porta la sbarra.

Example (1-1) can be understood in two completely different ways: *the old lady brings the bar* or *the old door bars her*. In the first interpretation, *la vecchia* (the old lady) counts as a noun and plays the role of the first actant, *porta* (brings) counts as the bivalent verb, *la sbarra* (the bar) again counts as a noun, playing the role of the second actant. In the second interpretation, *la vecchia porta* (the old door) is grouped as the first actant, while *sbarra* (bars) acts as the bivalent verb. Finally, the pronoun *la* (her) is a still unresolved anaphoric place-marker of a noun, playing the role of the second actant.

The two interpretations correspond to two similar but distinct adtrees [15], as shown in Figure 1.1. Both interpretations are fully acceptable as the ambiguity

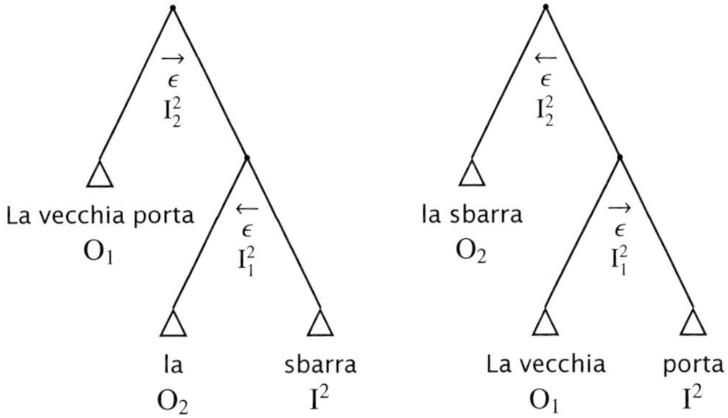

Figure 1.1: The two interpretations of *la vecchia porta la sbarra*

stands in the parser of the Italian language. In fact, two chains of constructions equally valid and semantically plausible can generate the respective adpositional trees. Also in English there are some known examples of sentence where there is more than one valid syntactic relationship, for example:

(1-2.) Time flies like an arrow and fruit flies like a banana.

where 'flies' counts as a verb in the first occurrence and as part of the noun in the second one.

By contrast, there are cases where there is only one adpositional tree but there is more than one semantic interpretation.

(1-3.) È la rossa di Maranello.

Example (1-3) can be interpreted as *She's the red-haired girl coming from Maranello*, an Italian town in the Emilia-Romagna region, as well as *it's the Ferrari car*, meaning the famous sport car produced in Maranello.

In this case, we have two different meanings for the same text, under the same syntactic interpretation: the adtree will be the same (Figure 1.2).

Adtrees are built from some basic bricks via a few standard constructions. Correctly identifying the bricks and the basic constructions is the main objective of this book. But, let's suppose to have them: then, we can generate the whole language and, vice versa, we can check if a piece of text is syntactically correct by asking if there is a corresponding adtree. Moreover, we would say that a piece of text is syntactically ambiguous if it admits more than one correct adtree. [16]

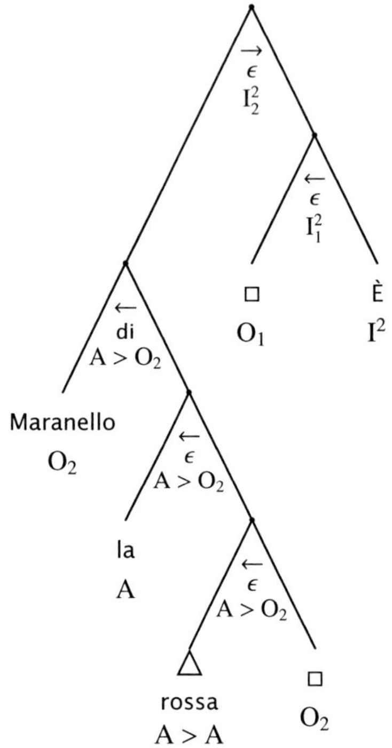

Figure 1.2: The adpositional tree of *È la rossa di Maranello*

In mathematical terms, adtrees and constructions between them form a structure **AdTree** which is a *category*[17], see Mac Lane (1998) and Borceux (1994). A mathematical category is an algebraic structure composed by two classes, the *objects* and the *arrows*; arrows lie between two objects, the *source* or *domain*, and the *target* or *codomain*. Also, a category states that there are distinct arrows, the *identities*, one for every object A and such that the source and the target are A. Moreover, a category is equipped with a partial operation allowing to compose two arrows whenever one has the domain which is the target of the other one. Composition is required to be associative and identities act as one expects with respect to composition.

Intuitively, there is an arrow f from A to B whenever we can construct the B tree starting from the A tree applying the construction f. We do allow complex constructions obtained by sequentially composing simpler ones; if f and g are constructions such that $f(A) = B$ and $g(B) = C$, that is, if f maps A into B, and

g constructs C from B, then $g \circ f$ is the construction which maps A into C by doing g after f.[18]

We observe that, calling M the free monoid over the alphabet of morphemes of some natural language, i.e., the set of all possible (finite) sequences of morphemes obtained by juxtaposition, the functions mapping the trees in **Adtree** into the sequences of M comprehend the textual renderings of adpositional trees. If we restrict our attention to *contravariant functors*, i.e., the functions preserving the identical transformation and the reverse composition of adpositional trees, we get a class of functions which is called *presheaves over* M. Requiring that a presheaf maps morphemes in the adtree into themselves in the monoid, we get exactly the lexicalizations of adtrees. In other words, there is a subclass of presheaves which directly corresponds to the texts the adtrees represent and which encodes the transformations that constitute the grammar. It is this space of presheaves which is generally understood as the subject of linguistics.

It is possible and fruitful to interpret the basic constructions of Category Theory, e.g., natural transformations, limits and colimits, etc., in the framework we have just introduced. But it requires a better understanding of the basic blocks of the **AdTree** category. So, we stop here in our analysis, at least until the following chapters will introduce the required elements.

As a side effect of this intended model of interpretation, it follows that whatever construction over adtrees which is built by combinatorially composing the fundamental constructions, is an arrow. Lifting the structure of the **AdTree** category into the spaces of presheaves, which is a category, we can reason in a larger and richer environment, where the full power of mathematical methods can be applied: in fact, the presheaves space is a *Grothendieck topos* (Mac Lane and Moerdijk, 1992; Johnstone, 2002), one of the richest mathematical structures we can deal with.[19]

As we have suggested, categorial mathematics give a set of elegant tools to build natural language grammars because it provides a transparent account of the correspondence between the syntactic and the semantic combinatorics—in other words, how meaning is reflected into collocation and word order phenomena, hypothesis put at first by Montague (1973) in Hintikka et al. (1974).

From a mathematical point of view, formalisms based on categories,[20] usually employ some variant of the Lambek calculus, because it corresponds to a non-commutative, substructural variant of linear logic. So, Lambek calculus, as well as derived formalisms, is based on a logical system where we have limited *resources*, allowing to model, e.g., that a phrase may contain at most one verb, and connectives are not commutative, modelling the ordering of elements in a phrase. An important result about pure categorial grammars is (Pentus, 1993) that shows how their generative power is that of context-free grammars and thus inadequate for theories of natural language syntax; this result applies only to categorial grammars which have a natural computational counterpart defined as an extension of the λ-calculus, i.e., where composition rules are functional applications. Unfortunately, leaving the simple and safe world of λ-abstraction and application in favour of more sophisticated rules allows for an explanation of more complex linguistic phenomena, but at the price of having models whose behaviour is far from being well-understood. In fact, they have been nearly abandoned for the automatic processing of natural languages.

A modern formalism currently under development is the Combinatory Categorial Grammar (CCG) by Steedman and Baldridge (2007), a lexicalized grammar formalism which can be parsed in non-deterministic polynomial time. It is especially used in the field of statistical parsing. Linguistic categories are assigned to words by the lexicon, e.g., an intransitive verb has the category $S\backslash NP$ while a transitive verb has the category $(S\backslash NP)/NP$. Under a theoretical linguistic point of view, a CCG is constituent-driven, i.e., it retains the advantages and limits of the Chomskyan syntactic perspective we put into question here, so we can't use CCGs as such. Under a mathematical point of view, a grammar is a set of inference rules controlled by the linguistic categories, interpreted as functional spaces, see van Benthem (1995); such rules can be inferred by suitable techniques from machine learning (Manning and Schütze, 1999), essentially using statistical measures. Also, the probabilistic approach is used to cope with the fact that non-deterministic polynomial problems cannot be efficiently solved, as far as modern mathematics knows.[21]

Ranta (2004) proposes a formal framework for writing grammars and linguistic theories—following the motto *sentences-as-proofs*. From a mathematical point of view, it is based on Martin Löf Type Theory, see (Martin Löf, 1984), using the Coquand's algorithm for type construction/partial proof derivation (Coquand, 1996). So, it is ultimately based on a dependent-type, predicative variant of the λ-calculus.

There are two advantages in this line of research: first, the Grammatical Framework is a piece of software which can be used freely and it has a growing up community (Ranta, 2009); second, it might be used for implementation regardless of the linguistic theory behind. However, software implementation is out of the scope of the present book, which is dedicated to linguistic analysis.

Dynamic Syntax is based on the program of Labelled Deductive Systems, which aims to bring semantics back into syntax (Kempson et al., 2001). Labelled formulae and deductive systems are connected with algebraic frames and type logical grammars. In fact, Labelled Deductive Systems have been first introduced by Gabbay (1996) as an algorithmic way to control proof development using labels from some simple algebra whose operations are used to model the potential application of inference rules.

Unlike Chomskyan grammars, in Dynamic Syntax the trees represent the structure of the sentences first as the result of a growth obtained by a parsing strings uttered in contexts—this is the sentence's 'dynamics'. As it will be clear in the sequel of this book, we follow the hypothesis posed by Dynamic Syntax that (morpho)syntax is the mechanism for constructing representations of content. However, there are important differences, namely the use of linguistic constructions—based on cognitive linguistics results—is formally resolved in a totally different way. Moreover, while Dynamic Syntax makes use of Hilbert's epsilon calculus in order to solve quantified noun phrases, our formal bases are completely different.

1.3 What is in this book?

The aim of the present work is foundational. In particular, we want to give the right hints to understand natural language grammars within the adpositional paradigm. For us, 'understanding' means both having insights about the functioning of real-world usage of a given natural language—in terms that are comparable with other languages as well—and having insights about what is structural in a given natural language, in constructive mathematical terms. In practice, there are no distinct "linguistic *vs.* mathematical" insights; rather, the joint linguistic and mathematical apparatus should be regarded as the two sides of the same coin.

Under a different perspective, this book is foundational as here the reader can acquire all the instruments necessary for the founding of Constructive Linguistics, i.e., an application of constructive mathematics to the realm of linguistics—broadly, the study of natural languages. When foundations are posed, many building details are still to be defined. In this book, we indicate which details, instantiations, specific phenomena are not covered, in order to help the reader see possible further directions of this work, standing on the solidity of the foundations posed here.

A crucial part of the foundations are the instruments from Category Theory, in particular, the intuitions about sheaves and Grothendieck topologies—the impatient reader is invited to read Appendix B at first.

The choice of the mathematical instruments apt to represent the results obtained within the adpositional paradigm was lead by two simple criteria: expressiveness and naturalness. From the point of view of expressiveness, it is easier to obtain more general result with an elegant formalism than narrower results with a less powerful formalism, and topos-theory is the strongest, general, elegant and broad formalism belonging to constructive mathematics we have nowadays.

On the other hand, the representation of linguistic phenomena in constructive mathematical terms should be natural. In other words, the mathematical constructs we use should have a direct and immediate counterpart in linguistic terms: in that way, readers interested only in the linguistic side can still follow the intuition of the model, without being forced to use abstract tools whose linguistic nature is not clear. Nevertheless, we invite even readers not used to mathematical formalisms to give a chance to our formal model, in reading the whole book in the order of presentation, even Appendix B. We are confident that—after the reading and understanding of the main text—the Appendix will be readable with profit and interest to all readers.

CHAPTER TWO

SYNTAX

How is a linguistic construction made? The constructive adpositional grammar approach gives an answer at more than one level. In this chapter, we will read the answer at the syntactic level. With 'syntax', we mean the *covert* structure of a construction, i.e., the relation between semantics and the overt syntactic stratum, which is reflected in the word order and collocation phenomena.[1] In other terms, our syntax model will clarify the association between meaning and grammar within the model.

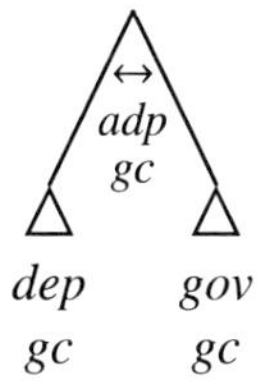

Figure 2.1: The abstract adpositional tree structure

The fundamental tenet of the adpositional paradigm is that grammar can be analysed into relations between each pair of linguistic elements, which are recursively defined. Figure 2.1 shows the generic adpositional tree—by now, *adtree*—where this tenet is made explicit.

Linguistically, the governor (*gov*) is the element of the triple—formed by governor, adposition, dependent—that can 'stand alone', i.e., without the dependent. The method we follow in order to analyse the data is *collocation*.

Within the cognitive perspective, languages are expressions of human cognition, therefore grammar is a conceptualisation principle, i.e., it is driven by the general rules of human cognition—in all its parts, i.e., phonology, morphology, syntax, semantics and pragmatics. What follows is that the collocation of words in the syntagmatic axis is what we should analyse to understand which construction is in action. Using the words of Taylor (2002, 11): "the very wording that we choose in order to linguistically encode a situation rests on the manner in which the situation has been mentally construed."

Hence, we take the same scene depicted through the analysis of *minimal pairs*: two chains of linguistic elements, identical except one single element. Then, we check the grammaticality on the syntagmatic axis. [2] If the grammaticality is not granted, then we know that the element is a governor. Otherwise, what changes in the reduced expression is the encoding of the situation. For example, in the construction *good corn sold here* we can cancel *good* preserving grammaticality, obtaining the reduced construction *corn sold here*, where the loss is in the fact that the corn is asserted to be good there. By contrast, if we cancel *corn* the result is not equally valid, i.e., *?good sold here* [3]. Therefore, *corn* is the governor while *good* is the dependent in their relation within the construction Y *sold here*, typically a written advertisement.

From a formal point of view, the analysis of minimal pairs can be understood as the fact that objects in a mathematical category are linked by arrows with a single source and a single target. Specifically, the source object is the governor, while the target object is the resulting adtree; the dependent chooses the specific arrow among the family denoted by the adposition. [4] So, grammatical constructions are loosely identified with families of arrows indexed by adtrees, where the index plays the role of dependent. In the previous example, the reader should notice how this interpretation strictly follows the pattern Y *sold here*, where Y now is really a variable standing for the index in a family of arrows.

The relation between a pair of governor-dependent is signed by the 'hook', i.e., the adposition (*adp*) put rightly under the root of the tree. The adposition marks the morpheme whose function is to govern the subordinate adtree, which stands on the left branch by convention—if there is no adposition, i.e, the relation is syntactic in nature, an epsilon (ϵ) will be put there. Furthermore, the adposition gives the *final* grammar character (*gc*) of its subordinate adtree (see next chapter for details). Another convention we adopt is that the governor (*gov*) always stands on the right branch of the current adtree, while the dependent (*dep*) always stands on the left one (see next sections for details). [5]

This structure is recursive, i.e., *dep* and *gov* can be adtrees. If empty, they are signed by a box ($\sqcap$). Please note that *adp* cannot be an adtree, i.e., it is always a morpheme, possibly empty (ϵ). The triangle ($\triangle$) signs that there is an adtree 'packed' in the linguistic expression. In other terms, the leaf, either governor (*gov*) or dependent (*dep*), brings *implicit* information that was hidden—essentially, for reading. This notational convention follows the constructive diktat of 'hiding information' in a way that can be recovered on need. In this respect, we follow Sambin and Valentini (1998) and the tradition of constructive and predicative mathematics.

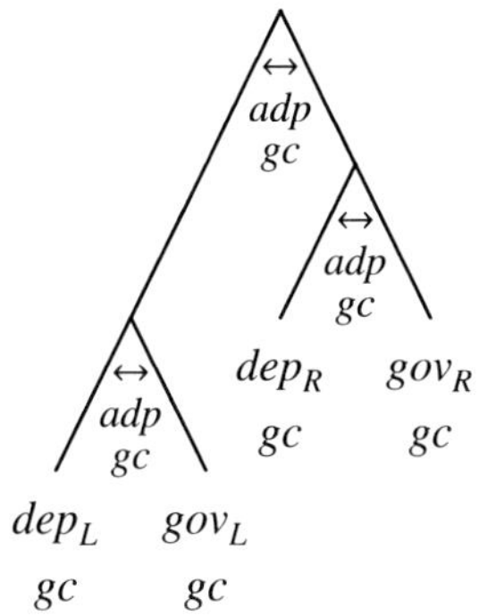

Figure 2.2: Example of recursive adpositional tree

Figure 2.2 shows exactly what just declared: the adtrees under the triangles of Figure 2.1 have been made explicit (dep_L and gov_L standing simply for ‘left dependent’ and ‘left governor’, while dep_R and gov_R stand for the right ones). To avoid tree branch clashing, we can lengthen the branches accordingly: please note that this fact is only for drawing needs, i.e., it is not a formal fact. By convention, when more than a governor (*gov*) is present as in Figure 2.2, the lengthened branch will be the left one. Otherwise, the lengthened branch will be the one on the right. [6]

It is worth noticing that the two branches are asymmetrical: this is not only a convention but it has precise theoretical reason. That is, a governor (*gov*) can have more than one dependent (*dep*) while the opposite is not possible. This is very important when dealing with verbal-based constructions (Section 2.3).

The consequence is quite obvious: the only possible extended adtrees are similar to the one shown in Figure 2.3, where the right branch governor (gov_R) rules over two dependents (dep_{R1}, dep_{R2}).

From a strictly formal point of view, the asymmetry can be interpreted as the fact that, although the involved adtrees (dep_{R1}, dep_{R2} and gov_R) are constructed independently from each other, the ‘rule in the grammar’ which allows the formation of the composed adtree, is strictly related to the governor and takes into account only the grammar characters of the dependents. In fact, in the formal model, we could say that there is an arrow from the object gov_R to the object depicted in Figure 2.3, while there is no arrow from dep_{R1} or dep_{R2} to the same object—and this is obvious since dep_{R1} and dep_{r2} are indexes in families of arrows. The reader is invited to notice that partial constructions are admitted: as in

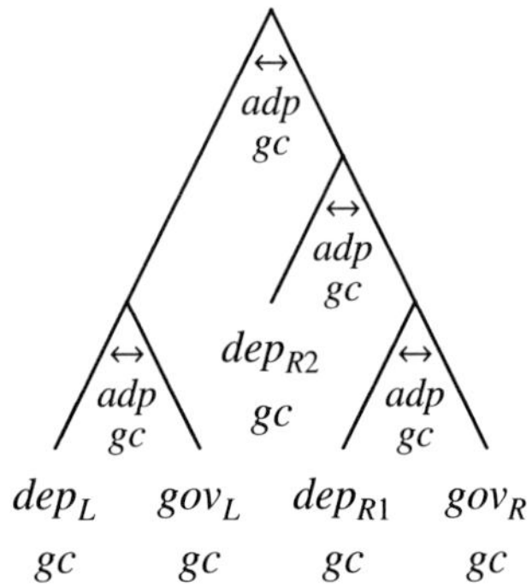

Figure 2.3: Example of an extended adpositional tree

Figure 2.4, if there is a construction leading from the leftmost adtree to the rightmost one, then there are two partial constructions which are the ones described by the ⇝ arrows.

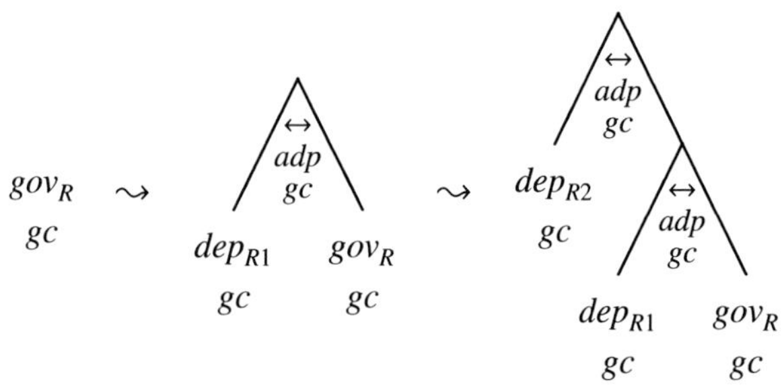

Figure 2.4: Example of partial construction

2.1 Government

Within the adpositional paradigm, the relation between the governor and the dependent is used to categorise the linguistic data. In fact, a speaker may choose either the governor or the dependent to be prominent; however, sometimes he cannot or do not want to. [7]

The *trajectory* is the direction of the information prominence and it is indicated by the adposition (*adp*). By convention, its start point indicates where

Table 2.1: Indicators of government and dependency

type of relation	*direction*	*indicator*
government	gov-to-dep	←
dependency	dep-to-gov	→
generic	both, irrelevant, or unknown	↔

the information prominence stands in that particular adtree, and it can be either the governor (*gov*) or the dependent (*dep*). Thus, the trajectory can indicate two possible directions, the one being the reverse of the other: from the governor to the dependent and from the dependent to the governor. We will refer to this kind of arrow—placed rightly above the adposition— with the term *indicator*, in order to avoid name clashing with other terms already familiar to the readers belonging to the mathematical communities. The bidirectional indicator (↔) is used for general, underspecified purposes, i.e., when the direction is either still undecided, unknown, or not important.

In sum, the possible indicators are three: either the trajectory goes in the right-left direction (←, government), or in a left-right one (→, dependency), or bidirectional (↔), as shown in Table 2.1.

We have a relation of *government* when the governor rules over the dependent because the governor is the most prominent element in the couple. It is the simplest case: typically, unmarked linguistic phenomena show government.

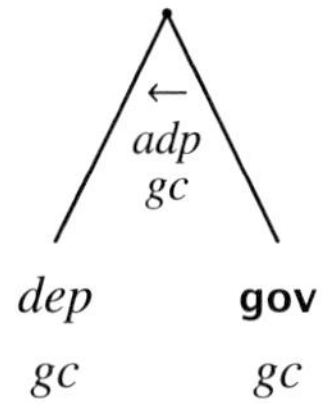

Figure 2.5: The abstract government relation

Figure 2.5 shows the abstract government relation: the trajectory starts from the governor (*gov*, here in bold for emphasis), by convention always put at the right branch—and ends into the dependent (*dep*).

Adtrees are recursively defined, and so trajectories—a structural part of any adtree. This formal fact has an exact linguistic counterpart: we can show the type of relation at different language levels, from phonology up to discourse pragmatics, covering all the linguistic spectrum—morphology, syntax and semantics included. An aim of this book is to show concretely how this fact operates. For the moment, let us show some examples to clarify this crucial point.

(2-1a.) *gov* $\leftrightarrow$ *dep*
I've not seen Liza. ϵ She was not with Paul.

(2-1b.) *gov* $\rightarrow$ *dep* $\leadsto$ *don't ask*
I've **not** seen Liza, ϵ she was not with Paul.

Examples (2-1ab) pertain to the same scene. For 'scene' we mean what depicted linguistically within the joint attentional frame, i.e., a perceptual situation shared by the participants. [8] The same scene can be rendered linguistically in very different ways: the trajectory gives an elegant and effective account to deal with very subtle differences in meaning otherwise neglected in other linguistic formalisms, as far as we know. [9]

In (2-1a) it is shown the simplest case, where no special stress, emphasis or other prosodic phenomena are used by the speaker. We assume that normally a phrase that follows another one, will be its dependent: in other words, what follows depends on what already put in the syntagmatic axis. The trajectory is bidirectional, as the information prominence is underspecified.

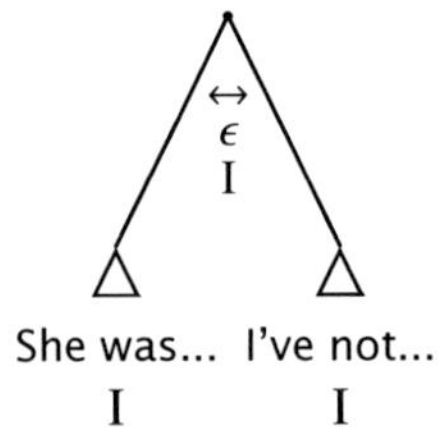

Figure 2.6: The packed tree of *I've not seen Liza. She was not with Paul.*

Figure 2.6 gives the representation of (2-1a) as an adtree, where each leaf is a complete phrase (I, as we will see in details in the next chapters) and therefore it is still not explicit here—this is why the triangles (△).

Example (2-1b) shows a complementary case, in particular a relation of government occurring within two distinct phrases. [10] The speaker has put the main stress on the word *not* (in bold to show the pragmatic emphasis), which gives a cue of the non-locutionary aspect of the linguistic act, put explicitly with the

Figure 2.7: The packed tree of *I've **not** seen Liza. She was not with Paul.*

directive *don't ask* (see Chapter 6 for more details). Figure 2.7 shows the corresponding adpositional tree.

2.2 Dependency

The trajectory of the information prominence can be reversed compared to government, i.e., the information prominence belonging to the dependent. This is the complementary case of government, and its name is *dependency*.[11]

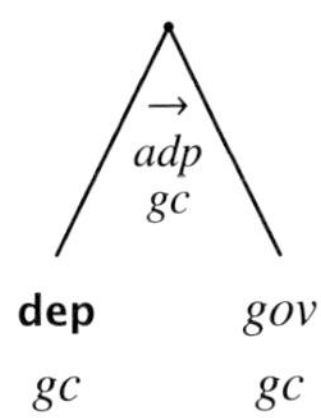

Figure 2.8: The abstract dependency relation

Figure 2.8 shows the abstract dependency relation (the dependent was put in bold for emphasis). Note that there is no other possibility of choice: either the trajectory corresponds to the governor, or to the dependent (or let underspecified, leading to the bidirectional relation, as shown in Table 2.1).

The following example pertain to the same scene seen above in (2-1ab), where by contrast the information prominence is completely reversed.

(2-1c.)	*gov*	←	*dep*	⇝	*you know*
	I've not seen Liza,		she was **not** with Paul.		*what I mean*

Example (2-1c) shows that the *second* occurrence of the word *not* (in bold) was stressed, and hence the non-locutionary information provided by the speaker is completely different, i.e., the fact that Liza was not with Paul, even if she was expected to be (put explicitly with *you know what I mean*).

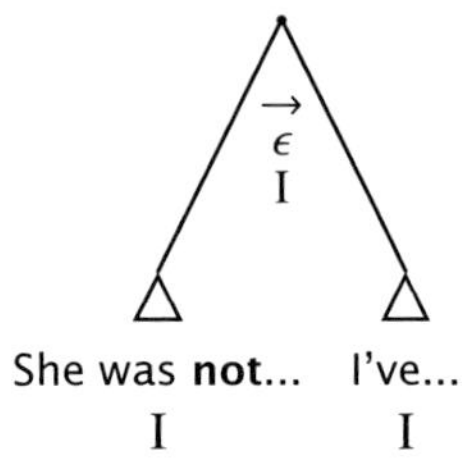

Figure 2.9: The packed tree of *I've not seen Liza. She was* ***not*** *with Paul.*

Figure 2.9 shows the correspondent adtree. It is worth noticing that the linguistic element which is different is only an emphasis, not even a morpheme. In fact, in all other aspects the wording of (2-1abc) are perfectly identical. This is in partial contrast to the quotation by Taylor (2002, 11) put above: his principle is fine, but it is not enough to give an account of the complexity of linguistic phenomena at a cognitive level. These minimal pairs show the expressive power—in linguistic terms—of the method of collocation applied within the framework of cognitive linguistics.

As a remark, we notice that government and dependency are not relevant for constructing adtrees in the formal model: in fact constructions are families of arrows denoted by adpositions and indexed by adtrees. But, indicators are extremely important because they govern the *transformations* between adtrees, that is, the way one can map an adtree to another one preserving the basic linguistic properties, such as their semantics. An in-depth treatment of transformations is postponed until the ingredients to properly model them will be introduced.

2.3 Valency

In an adpositional tree, the left and right branches follow different rules of construction. In particular, governors and dependents show an asymmetrical behaviour: a governor can have more dependents, while each dependent can have only one governor.

The number of dependents ruled by a governor is an important value because it gives the fundamental structure of the adtree. [12] We call this value *valency*. Following the definition by Tesnière, who first introduced it in linguistics (from chemistry), valency is a feature of the verb, and we will also use it mostly in relation with verbs:

> On peut ainsi comparer le verbe à une sorte d'**atome crochu** susceptible d'exercer son attraction sur un nombre plus ou moins élevé d'actants, selon qu'il comporte un nombre plus ou moins élevé de crochets pour les maintenir dans sa dépendance. (Tesnière, 1959, page 238, Chapter 97, par. 3, bold in the original). [13]

Even if we represent valency in a different way from the Tesnerian one, we still retain the use of the term *actant* to indicate the dependents belonging to a single governor. Hence, we will speak of first, second, third actant and so on (dep_1, dep_2, dep_3, etc.). By the way, it is worth remarking that, in constructive adpositional grammars, valency is a feature of the governor (*gov*)—not necessarily restricted to verbal entities. [14]

A governor has usually one, two or three dependents; the respective valency values are called *monovalency*, *bivalency* and *trivalency*. In trivalent constructions, usually an actant can be omitted safely, while the omission of two actants at the same time seldom occurs (see below for examples). In bivalent constructions, sometimes an actant can be omitted. However, it is not possible to state general rules as constructions are language-specific. It should be noticed that omitting an actant really means hiding it: although it is not written, it is 'present' in the construction, silently playing its role.

In rare cases a governor can have four dependents at the same time (*quadrivalency*, adjective 'tetravalent'), and in exceptionally rare cases five (*quinvalency*, adjective 'pentavalent'). A value of zero (*avalency* or also *zero-valency*) is also possible, and not so rare indeed: in this last case, the adtree collapses and reduces to a simple leaf/root.

Yes!

I

Figure 2.10: The collapsed tree of *Yes!*

Figure 2.10 shows an avalent governor of a simple standing alone, monomorphemic, phrase: *yes!*. However, when the valency value is strictly more than zero, the governor-dependent relations occur, in different combinations of governments and dependencies, according to the trajectory.

Consider the following examples of bivalent verbs (indicated by I^2):

(2-2a.) The bicycle belongs to Kim.

(2-2b.) Kim owns the bicycle.

Examples (2-2ab) are a minimal pair at a conceptual level, as they pertain the same scene, but the constructions activated by the English stems *belong* and *own* are different: in both cases, it is Kim who brings the information prominence, being the bicycle owner, even if it is not the first actant. The trajectory is carved into the construction as well as the valency value. [15]

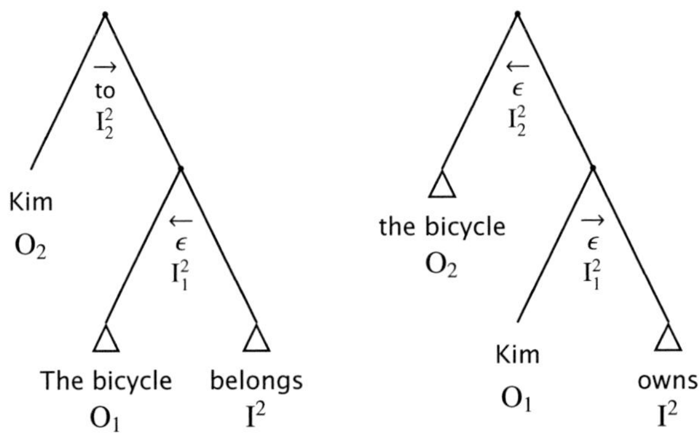

Figure 2.11: Two different bivalent trees

In the construction activated by the stem *belong* (Figure 2.11, left), the first actant (O_1) is prototypically inanimate (*The bicycle*). We say that the first actant has *saturated* the first valency of the bivalent verb *belongs* (I^2_1). Moreover, the relation activated by the first actant in this particular construction is of dependency (←), as this actant receives 'passively' the act of belonging. [16]

On the contrary, the second actant (O_2) which saturates the second valency (I^2_2) is primary responsible of the act of belonging, hence the relation will be of government (→). Similarly, the construction activated by *owns* shows a complementary arrangement of information prominence (Figure 2.11, right).

2.4 Transformation

The criteria to determine the type of the governor-dependent relation are by no means general but, on the contrary, language-defined, or, even more precisely, construction-driven. In other words, when a construction—i.e., a pattern of usage—emerges by frequency of use, the structure of the construction (adtree) becomes partially fixed in some linguistic elements. Sometimes even leaves can be fixed (as in the construction Y *sold here*, already seen at the beginning of this chapter through its instantiation *good corn sold here*), but, in every case, the *structure* of the adtree should be determined—i.e., the valency and the trajectory of the information prominence (as in Y *belongs to X* and *X owns Y*, just seen). [17]

In fact, the examples above should have shown that valency is an important variable, but it can be misleading without cues about the information prominence. It is also important to note by now that constructions emerge in order to establish a relatively coherent communication function in time, according to the existence of other constructions already established. In other words, within a single language its constructions constitute a system, that we call *construction system*. Therefore, we cannot observe a single construction in isolation but with its siblings—i.e., similar constructions, in synchronic and diachronic terms (again, we will apply the method of collocation, examining minimal pairs). In fact, even if synchronically the system—i.e., a living language in a given moment—runs perfectly, and hence there is no reason to change, languages changes constantly: this 'paradox of change', as put by the Romanian linguist Eugenio Coseriu, is the driving force of *grammaticalization*.

Grammaticalization is the force that explains the form of a construction diachronically, the construction system being its synchronic snapshot. In order to explain the inner form of the construction system we will use as the main linguistic tools of analysis the following: valency; trajectory of information prominence; transference (i.e., grammar character change obtained by adtree transformation, see next chapters for details). From a formal point of view, grammaticalization (diachronically) and the language system (synchronically) act in the same way: they *transform* a construction A to a construction B (in symbols: $A \rightsquigarrow B$) by modifying the tree structure of A in the tree structure of B.

In order to understand how transformation works, let us take a very famous example—slightly adapted—from a classic paper by Charles J. Fillmore (2003c), who, in the late 1960s, started to investigate the linguistic concept of 'case', also referring to Tesnière (1959). [18]

(2-3a.) The janitor will open the door with the key.

(2-3b.) The janitor will open the door.

(2-3c.) The key will open the door.

(2-3d.) The door will open with the key.

(2-3e.) The door will be opened with the key.

(2-3f.) The door will be opened by the janitor.

The English stem *open* is trivalent as it can give place to constructions with three actants, as in (2-3a). The *internal* actants (or: *in-valent* actants), in other words the ones pertaining to valency, are determined by the constructions they take place in. The criterion to determine if an actant is internal or external to valency (or: *extra-valent* actants) is the possibility of *advancement*: for instance, when an actant in the third valency in a fully explicit construction—here, only (2-3a)—can be advanced to the first position in a reduced construction (2-3c). It is important to note that *the scene should bc the same* both in thc fully explicit construction (2-3a) and in its reductions (2-3bf), even if the information prominence is different.

In order to understand this crucial point, we proceed by contradiction. Let suppose there exists a fourth actant (W) for the stem *open*. Let this fourth actant bring the semantic role of Benefactive, as in the following example:

(2-3g.) The janitor will open the door with this key for the Lady.

In other words, we are trying to verify if the construction is of the tetravalent type X *open* Y *with* Z *for* W instead of the trivalent type X *open* Y *with* Z.

(2-3h.) ?The Lady will open the door.

(2-3i.) ?The door will open with the Lady.

(2-3j.) ?The key will open the Lady.

Examples (2-3hj) show that no advancement is possible, therefore the supposed fourth actant Benefactive is extra-valent.[19] On the contrary, the Opener (X, *the janitor*), the Opened Objective (Y, *the door*) and the Instrument (Z, *the key*) can be advanced, hence they are in-valent and so the construction will have the following abstract, general form, which is trivalent: X *open* Y *with* Z.

Now, it is worth recalling here that we consider semantic roles an open inventory, not a fixed list, because semantic roles belong to the semantic frames, which are language-defined.[20] In this respect, we consider the empirical treatment of linguistic data by most cognitive linguists convincing.

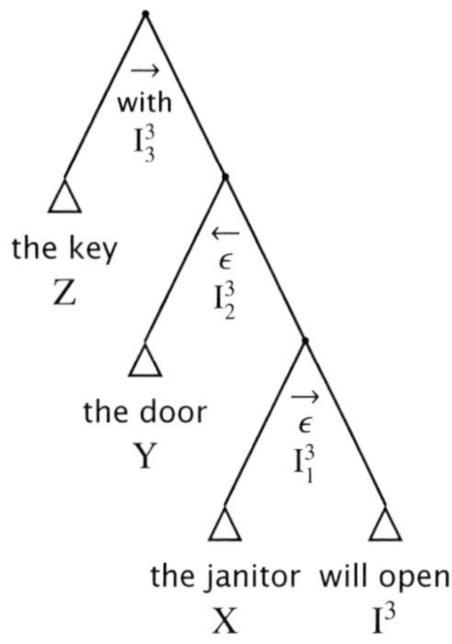

Figure 2.12: The tree of *X open Y with Z*

Figure 2.12 shows a rather complex adpositional tree, which represents the construction of (2-3a). We can notice that the first and third actants (X and Z, *the janitor* and *the key*, respectively) govern their relations as they 'do something', while the second actant (Y, *the door*) is in a relation of dependency.[21]

Now, it is interesting to see how (2-3bf) are *reductions* of the fully explicit construction (2-3a). We call them *reduced constructions* as (a) some actants are omitted and (b) some other actants are advanced, according to needs.

In particular, (2-3b) is a trivalent verb (*open*) used as if it were bivalent by omitting the third actant (*with the key*): no advancement is needed (Figure 2.13).

Examples (2-3cd) show a similar case: *the janitor* is omitted, the rest is advanced by one position (Figure 2.14).

On the contrary, (2-3c) shows both an omission and an advancement (Figure 2.15, left): initially, the first actant *the janitor* is omitted; after the omission, the first valency (X) is let free and hence the third actant *with the key* can be advanced to the first position (Z $\rightsquigarrow$ X), with a further omission of the preposition marking the INSTRUMENTAL. The resulting tree is shown in Figure 2.15 (right).

Finally, examples (2-3ef) show two possible English passive constructions generated by the stem *open*, Y BE *opened by* X (2-3e) and Y BE *opened with* Z (2-

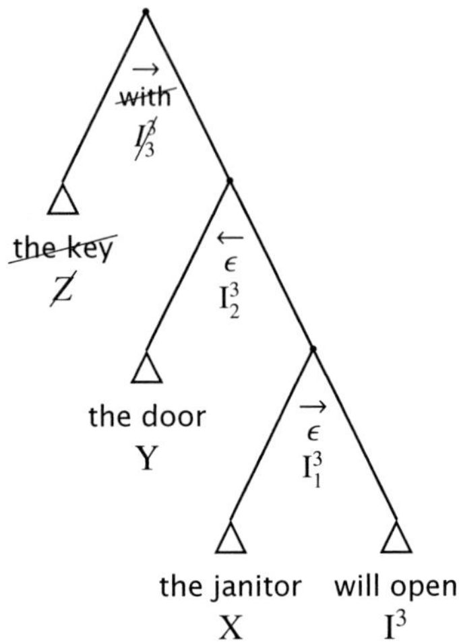

Figure 2.13: The reduced tree of *The janitor will open the door*

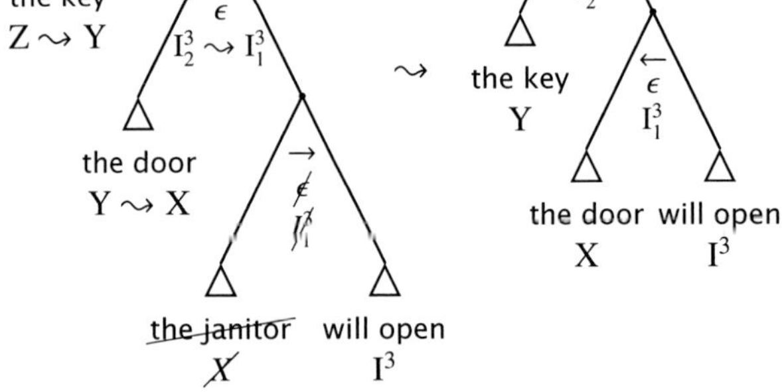

Figure 2.14: The tree of *The door will open with the key*

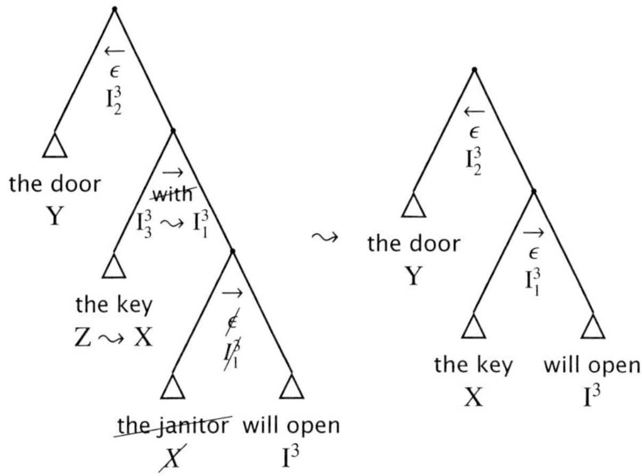

Figure 2.15: The tree of *The key will open the door*

3f), where the first valency is occupied by the originally second actant (Y, e.g., *the door*). It is worth noticing that every English passive construction transforms a trivalent verb in bivalent ($I^{3\rightsquigarrow 2}$), besides the changing in the pattern of the information prominence.

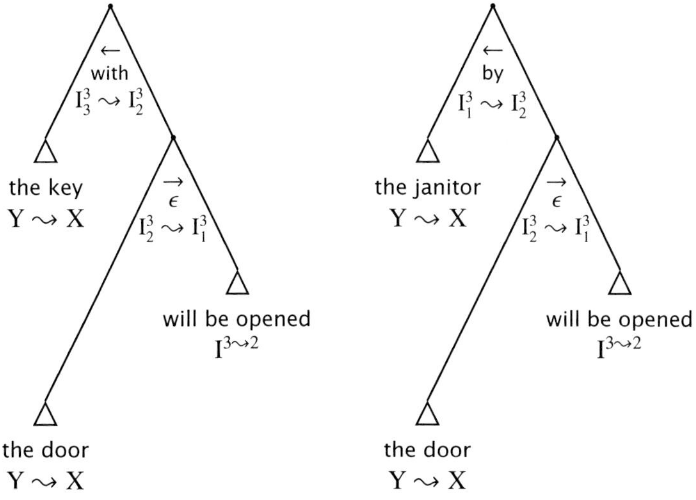

Figure 2.16: The trees of *The door will be opened...*

Example (2-3f, Figure 2.16, right) shows the extraction of the first actant (X) as the most prominent information in the phrase, while (2-3e) is simpler, as it simply omits the first actant, while the others are advanced by one position (Figure 2.16, left: the left branch of the adpositional tree belonging to *the door* was lengthened only for clarity).

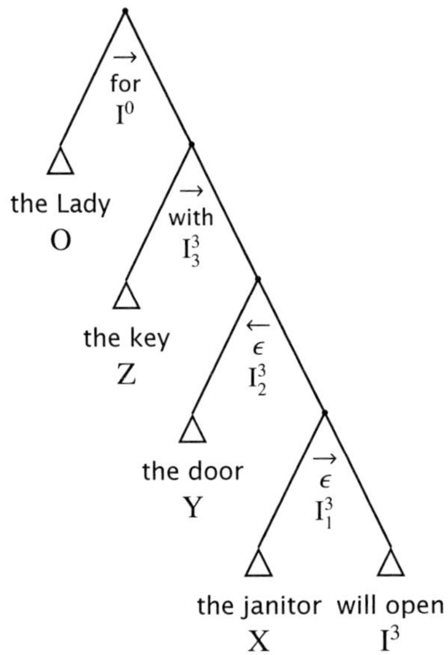

Figure 2.17: The tree of *The janitor will open the door with the key for the Lady*

From a strictly formal point of view, we are saying that adtrees are 'generated' by the valence (and, possibly, other construction rules), and then transformed to their final form. This separation between generation and presentation of an adtree is important to support the constructive character of our approach, as transformations are responsible for the hiding of information, which is present in the generation phase.[22] For instance, Figure 2.17 shows the generated adtree for example (2-3g). Each of the examples (2-3bf) is obtained by hiding some pieces of this adtree, which are barred in the derived adtrees. Usually, the barred pieces are omitted, as in Figure 2.13, or written as ϵ, the zero morpheme, when inside the adtree, as in Figure 2.15. Although this convention is useful, it may be misleading, since the pieces are not absent, but really hidden, i.e., they play a role when we want to apply further transformations to the adtree. This fact should be evident since transformations are first composed, and then applied to the generated germ.

A case study: the dative alternation construction

Constructive adpositional grammars give a simple and effective explanation of debated nuances in natural language grammars. For example, let us take the English construction known as 'dative alternation' (or, in less recent times, 'dative movement'), which shows how grammaticalization features as the nominal *vs.* pronominal marking of an actant can activate different constructions. Dative alternation is a well-studied phenomenon especially in English, but also in other languages, such as German, Dutch, Brazilian Portuguese, and Tsunga (a variety of Shona, a Bantu language). [23]

According to what previously said, each different construction corresponds to a different collocation, and this fact is structurally reflected in a different pattern of trajectories representing the information prominence.

(2-4a.) Ann gave Kim the car. (not the bicycle)

(2-4b.) Ann gave the car to Kim. (not to Bruce)

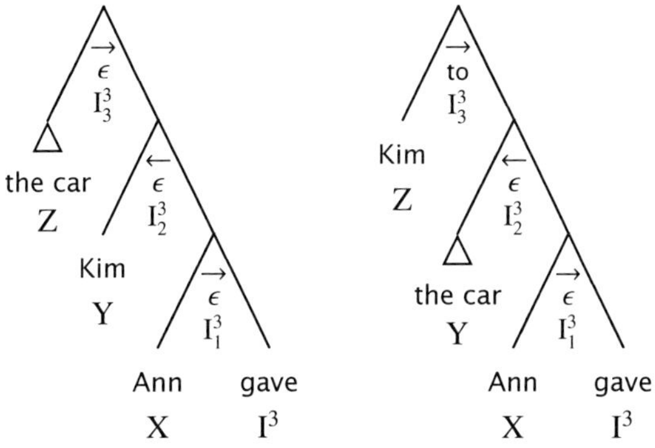

Figure 2.18: The trees of *A. gave K. the car* and *A. gave the car to K.*

These two constructions (Figure 2.18) are different, as in both cases it is the *third* actant that is put in evidence. In fact, (2-4a, Figure 2.18, left) shows that the most salient information—in other words, the phrasal trajectory—is *the car*, while by contrast in (2-4b, Figure 2.18, right) it is *Kim* the most prominent element within the phrase.

(2-4c.) Ann gave him the car.

(2-4d.) *Ann gave the car to him.

This analysis is confirmed by the fact that (2-4c), which resembles (2-4a), is perfectly grammatical, while (2-4d) is hardly acceptable. [24]

According to Abreu Gomes (2003), the Dative alternation construction in Brazilian Portuguese is similar to (2-4b).

(2-5a.) Maria deu o livro para/a/ϵ Pedro
Mary gave the book to Peter

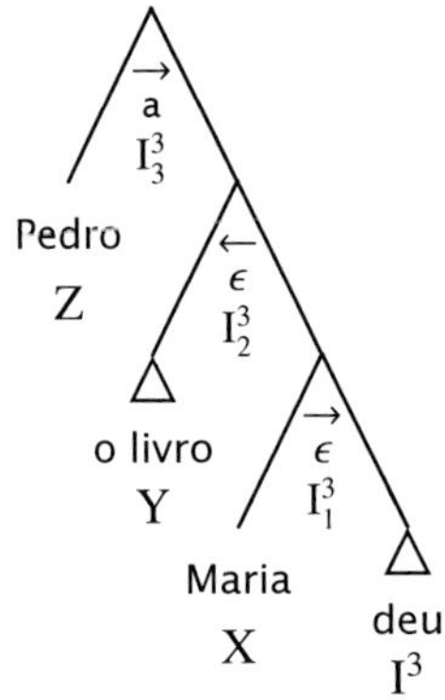

Figure 2.19: The tree of *Maria deu o livro a Pedro*

In (2-5a), the preposition introducing the explicit nominal Receiver (in English invariably marked by the preposition *to*) can be rendered with *para*, *a* or even a zero morpheme (ϵ), according to different degrees of grammaticalization.

In Italian, the pronominalisation of the Receiver invokes a construction similar to the English one with a zero marker (2-4c), while the presence of an explicit nominal marker activates a construction similar to the English one presenting the preposition *to* (2-4d).

(2-5b.) Maria diede il libro a Pietro
Mary gave the book to Peter

(2-5c.) Maria gli diede il libro
Mary him gave the book

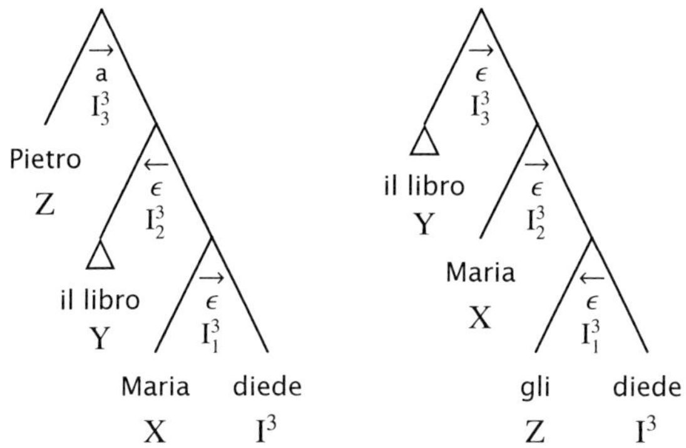

Figure 2.20: The trees of *M. diede il libro a P.* and *M. gli diede il libro*

Figure 2.20 (right) shows the tree describing (2-5b), where the third actant is advanced (meaning the exact sense already explained) next to the verb *diede*, *gli* being a clitic, which is advanced next to the verb[25].

Determining advancements is made through the comparison of construction siblings. It is obvious that the instance in (2-5b), showing explicitly the Receiver, and the instance in (2-5c), showing it implicitly via a clitic, are strongly related. Furthermore, the pronominal Receiver can be safely omitted (2-5d), while the second actant, the Objective, is almost mandatory (2-5e).

(2-5d.) Maria ~~gli~~ diede il libro
Mary ~~him~~ gave the book

(2-5e.) ?Maria gli diede ~~il libro~~
?Mary him gave ~~the book~~

This minimal pair shows how usage-based language data is used to determine the structure of a construction in terms of valency and information prominence.

It is worth noticing that there are some languages that encode the information prominence into the construction with case marking, for example German.

(2-6a.) Der Mann gab der Frau das Auto
The man/Nom *gave the Lady*/Dat *the car*/Acc

(2-6b.) Der Mann gab es der Frau
The man/Nom *gave it*/Acc *the Lady*/Dat

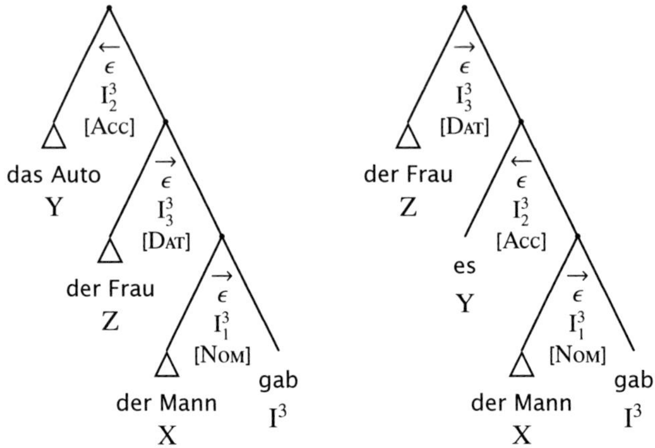

Figure 2.21: The trees of *Der Mann gab...*

In (2-6a, Figure 2.21, left), the third actant is nominal (*der Frau*) and next to the verb (*gab*), while in (2-6b, Figure 2.21, right) the second actant is pronominal (*es*) and it shifts in the second position. In this respect, German behaves exactly the opposite to Italian (see the various instances in 2-5). [26]

Sometimes there are two competing constructions with the same morphology but different collocation, as in Tsunga—reported by Abreu Gomes (2003).

(2-7a.) Wakonge wapele mbyia wanace
Woman gave money children

(2-7b.) Wakonge wapele wanace mbyia
Woman gave children money

Here, the information prominence is left underspecified (↔) because more actual data of construction siblings should be provided to identify information prominence, if any. In fact, it is possible that in Tsunga there is no prominence in this minimal pair of constructions.

In sum, in this chapter we have seen what are the main components of the syntax of adpositional trees, i.e. adpositions, signing the governor-dependent organisation as well as the trajectory of the information prominence, together with the structure of the construction tree skeleton, derived from the valency value of the main governor active within the phrase.

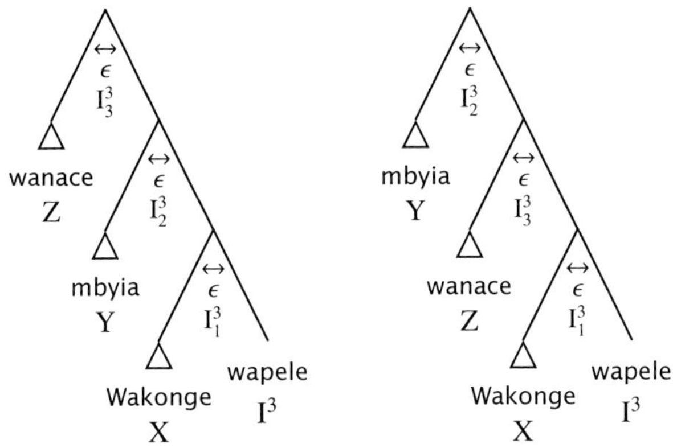

Figure 2.22: The trees of *Wakonge wapele...*

In the next chapter the picture will be completed at a intra-phrasal level, i.e., how a *single* phrase is made. The key concept will be the grammar character, which will help us to understand the inner dynamics of morphosyntax and the structure of the lexicon. We will reprise and define with more details every concept presented in this chapter, both in a formal way as well as through usage-based linguistic examples.

Chapter Three

Constructions

In order to understand how the construction of a single phrase is made, we have to introduce the third linguistic tool used within the adpositional paradigm, which is the *grammar character*. From now on, we take valency and information prominence as granted (see again Section 2.4 if needed). However, before to deal with grammar characters, we should explain how we consider the material on the syntagmatic axis which forms constructions at work.

In our perspective, the ultimate unit of linguistic analysis is not the word, but the *morpheme*. We use the word 'morpheme' in the standard sense, meaning a word—or a word part—which cannot be further analysed. For instance, in English *fretless* is a word made by two morphemes, i.e., *fret-* and *-less* (the minus indicates that they are bound one with each other), while *girl* is a word made by one morpheme only. [1] Morphemes work together to form linguistic constructions, in the sense put by cognitive linguists—an example of construction being Y BE *eaten by* X (Croft, 2009).

This perspective is quite unpopular among linguists in general and computational linguists in particular, who usually take as the basic unit of the analysis the notion of 'word'. [2]

In our understanding of the literature, this happens because the rough notion of 'word' is a string of characters belonging to the alphabet allowed by a particular language—written between two spaces (within a phrase). This rough definition—sometimes referred as the notion of 'words-as-strings'—seems to be intuitive to most speakers of Western languages and comfortable to natural language processing treatment.

Nonetheless, the advantages of this perspective, which appear clear at a first glance, come at dramatic costs. The first remark to be done comes from typology: not every language of the world having a proper writing system can be easily described in terms of words-as-strings—e.g., Chinese. The second remark is that spaces between words seem to be equal one with each other but they are not at all; for example, in English, the spaces between the following pairs of words are different: *per se*, *morning star*, *birds fly*, *hot dog*. We claim that most of the problems of (computational) linguistics nowadays are derived from the notion of words-as-strings: idioms, expressions, lexical intricacies are treated as

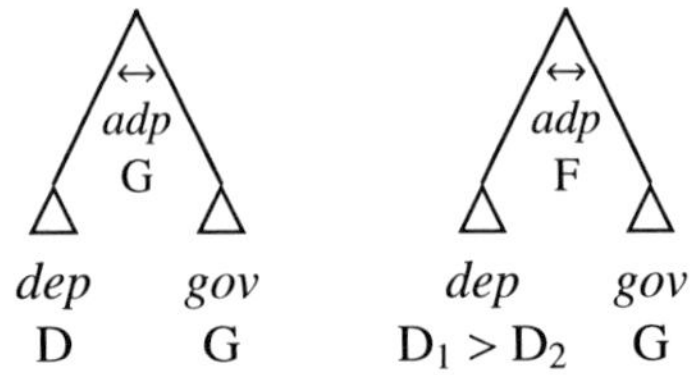

Figure 3.1: Abstract trees and grammar characters

a world apart, instead of being natural phenomena to be analysed in a standard way, i.e., as instances of one single model that encompasses and explains the main areas of linguistics: morphology (rules for morphemes), syntax (rules for words), phraseology (rules for expressions).

Under a linguistic perspective, we choose the notion of construction in order to give an account of the main areas of linguistics throughout a single concept. A construction can be seen as an adpositional tree pattern, where the final leaves—i.e., the single morphemes—belong to a particular class.[3] Under a mathematical perspective, constructions are families of arrows in a category, where the adtrees are its objects (see Appendix B for the mathematical treatment).

If a class is open and large, the construction will be productive; otherwise, if a class is closed and small—fixed, carved in the construction, at the limit—the construction will be idiomatic. Within the adpositional paradigm, a word is a particular construction—think for example to the English construction of stative plurals: 'add a *-s* to the stative morpheme, if it has the following features (phonotactic, morphological, etc.)'. In sum, there is no special status of words *per se*. The notions of morpheme and construction are linguistic universals: every language is, more or less, grammaticalized—otherwise there will be no regularity in morphemes into the syntagmatic axis. Hence, grammar is made by morphemes (the units) and constructions (their established patterns of usage). In this chapter, we will explain the morphology of constructions in general, while the morphology of words is derived as a special case in the next chapters.

We state that every element in the triple of the adtree—i.e., adposition, dependent, governor—has a definite grammar character (Figure 3.1), which can be considered by the moment as a generic kind of label (see below for details). The adposition is the marker of the grammar character of the whole adtree. Two scenarios are possible to determine the final grammar character of the adtree.

In the first scenario, the adtree inherits the grammar character from the governor (*gov*), as it 'governs' the adtree (Figure 3.1, left). Let D be the generic grammar character of the dependent and G be the generic grammar character of the governor. If the governor gives its grammar character to the whole adtree, we will say that the grammar character *raises* or that there was a *character raising*.

In the second scenario, the governor does not concur to determine the final grammar character. Instead, the final grammar character is imposed by the adposition, whose main role is exactly to change the grammar character of the governor, and sometimes (e.g., 3-11c) even the adposition of the dependent (Figure 3.1, right). If the grammar character of the whole adtree is imposed by the adposition, we will say that there was a *character imposition*. It is worth noticing, that the adposition imposes the trajectory of information prominence as well.

In a formal way, every adtree has a grammar character, or, equivalently, every adtree belongs to a class denoted by the grammar character. Any construction takes some adtrees—the parameters—and produces a new adtree: the parameter adtrees are supposed to be members of appropriate classes, i.e., they must possess the right grammar character.

For example, the first actant of a monovalent verb must be a stative, i.e., the first actant must be an adtree in the O class. The result of a construction is an adtree whose grammar character is either the one of the governor, which is the source object in the mathematical category of adtrees, or it depends on the particular construction.

As constructions are denoted by adpositions, a character raising denotes a construction which is 'generic', i.e., it decorates the governor with the dependents, that is, there is an arrow from the governor to the constructed adtree whose indexes are the dependents, and it has the indicator as its label; on the contrary, a character imposition denotes a construction which is devoted to coordinate the governor with the dependents to build a new adtree, possibly of a different class.

As a matter of fact, as we will show in the following, the grammar character of an adtree obtained by character imposition is fixed, i.e., it depends exclusively from the adposition and not from the governor or the dependents. In particular, it never happens that the dependent, as its name suggests, gives its grammar character to the adtree.

3.1 Grammar characters

There is a general agreement among linguists that the presence of expressions of reference (i.e., things) and the presence of predication (i.e., events) are 'conceptual archetypes' (Langacker), i.e., they are always-valid universals of language. In other words, they pertain to cognition and communication, as put by Tomasello, i.e., not to some languages in particular (Langacker, 1987; Tomasello, 2003, for example). Of course, things and events are shaped by languages in very different ways, and the traditional grammar categories of 'noun' and 'verb'—representing things and events respectively—are bound too closely to the Standard Average European Sprachbund (Haspelmath, 2001).

Hence, we decided not to use them in a general, abstract way, in order to avoid confusion.[4] In an almost neglected article by Whorf (1945), published posthumously, he addresses exactly this problem, comparing Western languages, such as English, with Non-Western ones, such as Hopi or Yana.[5] The main difference is that we talk about 'grammar *characters*' instead of 'grammar *categories*', as the term 'character' refers to a property, which implies that it can be present or absent. This is a better wording than 'category', which, in linguistics, usually reminds something fixed, i.e., a morpheme *always* belonging to a particular category (this particular case is called 'selection'; see below for details). Moreover, in this way we avoid name clashing with the mathematical usage of the word 'category' adopted here, as already said in the Introduction. For the same purposes, we will call the grammar character belonging to the expressions of reference (i.e., things) *stative*, while the one belonging to predication (i.e., events) will be called *verbant*.[6] Hence, the process of appending the property to the right morphemes will be called, respectively, *stativation* and *verbification*.

However, natural languages would be very poor if they would not find a manner to *modify* stative and verbant groups. The result of these modifications are respectively *adjunctives*[7] and *circumstantials*.

It is worth noticing that the familiar 'part-of-speech' are nothing else than *instantiations* of these grammar characters. Their taxonomies depend on the cultural traditions of the grammarians who wrote the grammars of the languages themselves. For example, in English linguistics, verbs as well as interjections are verbants (I), while name-entities, nouns and pronouns are statives (O). Moreover, determiners, articles and all adjectives are adjunctives (A), while adverbs and adverbials—i.e., groups of morphemes acting as an adverb—are circumstantials (E). By contrast, in Chinese linguistics coverbs count as adpositions (*adp*). The classes belonging to the grammar characters are language-driven.

Table 3.1 shows the four morphologic grammar characters. In the text, we will refer to them through their names or processes, while in the adpositional trees we will use the symbols (as already done in the previous pages), following the proposal by Tesnière (1959), who has used these grammar characters inde-

Table 3.1: The morphologic grammar characters

name	*process*	*role*	*symbol*
stative	stativation	fundamental character	O
verbant	verbification	fundamental character	I
adjunctive	adjunctivation	modifier of O	A
circumstantial	circumstantiation	modifier of I	E
adposition	adpositional	structural character	U

pendently by Whorf (1945). But before to present the properties of grammar characters in detail, we should state the relations that can occur between them in terms of adpositional trees, specifying what stated in general in Figure 3.1, with a special attention to the notion of valency already seen in Chapter 2.

Table 3.1 displays a fifth grammar character, the one of adpositions. This grammar character contains the morphemes which are used to link the other grammar characters into complex structures that are responsible for constructing linguistic expressions. In a way, it is easier to consider adpositions as members of a grammar character, denoted by the U symbol, as this assumption simplifies their formal treatment; but, this class of morphemes is 'different' as its role is to generate the structure of the language rather than to convey meaning—the reader can think of 'U' as standing for 'union'.

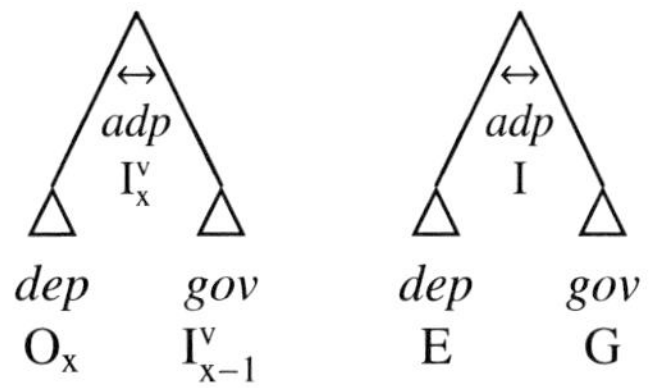

Figure 3.2: Abstract verbant relations

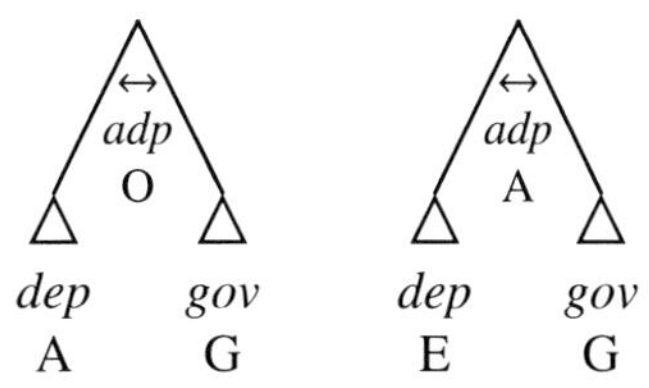

Figure 3.3: Abstract adjunctive relations

Figures 3.2 and 3.3 show the admissible adtrees in terms of relations between different grammar characters. Apart of the stative-verbant relations (Figure 3.2, left)—which are special because of valency—in all the other relations the grammar character of the governor is generic (G), i.e., either the grammar character will be *raised* from the governor, or it will be *imposed* by the adposition. [8]

It is worth noticing that the generic indicator (denoted by: ↔) can be specified both as a government (←) and as a dependency (→). So, as three instances for each adtree are possible, the abstract adtrees are exactly twelve. They are the following, in linear format: [9]

1. (relation stative-verbant) $adp_{\mathrm{I}}^{\leftrightarrow}((dep)_{\mathrm{O}}, (gov)_{\mathrm{I}})$
2. (government stative-verbant) $adp_{\mathrm{I}}^{\leftarrow}((dep)_{\mathrm{O}}, (gov)_{\mathrm{I}})$
3. (dependency stative-verbant) $adp_{\mathrm{I}}^{\rightarrow}((dep)_{\mathrm{O}}, (gov)_{\mathrm{I}})$
4. (relation adjunctive-stative) $adp_{\mathrm{O}}^{\leftrightarrow}((dep)_{\mathrm{A}}, (gov)_{\mathrm{G}})$
5. (government adjunctive-stative) $adp_{\mathrm{O}}^{\leftarrow}((dep)_{\mathrm{A}}, (gov)_{\mathrm{G}})$
6. (dependency adjunctive-stative) $adp_{\mathrm{O}}^{\rightarrow}((dep)_{\mathrm{E}}, (gov)_{\mathrm{G}})$
7. (relation circumstantial-verbant) $adp_{\mathrm{I}}^{\leftrightarrow}((dep)_{\mathrm{E}}, (gov)_{\mathrm{G}})$
8. (government circumstantial-verbant) $adp_{\mathrm{I}}^{\leftarrow}((dep)_{\mathrm{E}}, (gov)_{\mathrm{G}})$
9. (dependency circumstantial-verbant) $adp_{\mathrm{I}}^{\rightarrow}((dep)_{\mathrm{E}}, (gov)_{\mathrm{G}})$
10. (relation circumstantial-adjunctive) $adp_{\mathrm{A}}^{\leftrightarrow}((dep)_{\mathrm{E}}, (gov)_{\mathrm{G}})$
11. (government circumstantial-adjunctive) $adp_{\mathrm{A}}^{\leftarrow}((dep)_{\mathrm{E}}, (gov)_{\mathrm{G}})$
12. (dependency circumstantial-adjunctive) $adp_{\mathrm{A}}^{\rightarrow}((dep)_{\mathrm{E}}, (gov)_{\mathrm{G}})$

This list would be not complete without mentioning the identity relations, i.e. when the governor and the dependent have the same grammar character, and thus the resulting final grammar character of the adtree will be again identical.

Figure 3.4 shows the quite obvious adtree of identity relations: the most prominent element is the governor by definition, because it can 'stand alone' (see again Chapter 2 if needed). As we will see dealing with transference (next chapters), identity relations are the structural description of a very important universal property of languages: redundancy (Chiari, 2002).

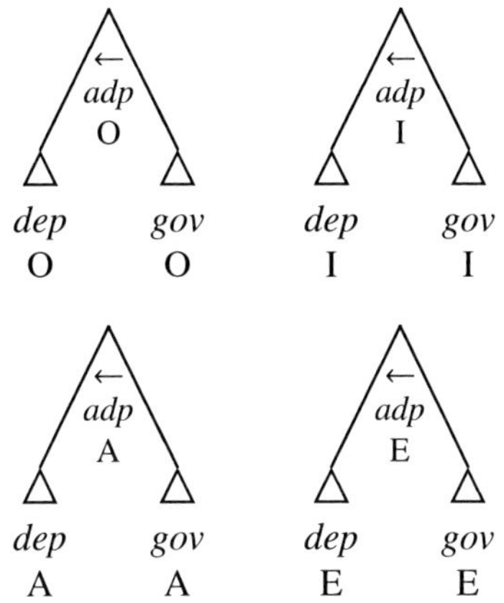

Figure 3.4: Abstract identity relations (redundancy)

3.2 The stative-verbant relations

The stative-verbant relations are the skeleton of most constructions. Within a single language, constructions are often built around a verbant group, eventually transformed by grammaticalization in something else, e.g., a stative group – a phenomenon called *normalisation* (see below).

This kind of relations is ruled by valency, as explained in the previous chapter (Section 2.3). Note that valency rules both verbal and interjectional groups. For instance, consider the following interjectional groups:

(3-1.) Ah!

(3-2.) Dear me!

Ah!

I

Figure 3.5: The collapsed tree of *Ah!*

Figure 3.5 shows that (3-1) has a valency value of zero, so the adtree collapses in a simple leaf.

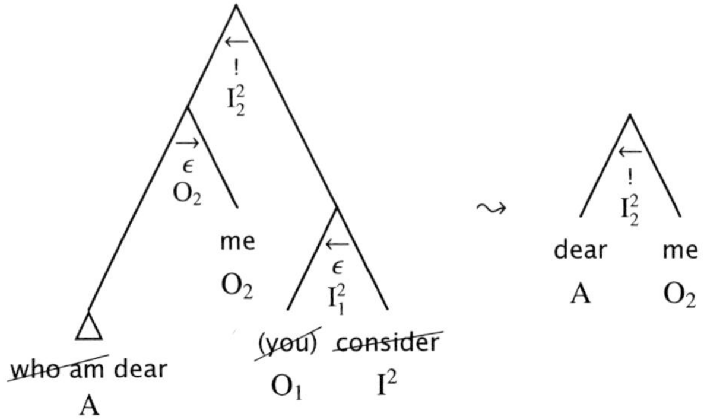

Figure 3.6: The explicit and reduced tree of *Dear me!*

By contrast, Figure 3.6 shows a *bivalent* interjection (3-2). In fact, the construction *dear Y!* can be instanced by oblique pronouns, such as *me*, *him*, *her*. This means that a bivalent verb was cancelled by grammaticalization, but a trace still remains because of the fossil oblique case. This kind of construction, usually referred to as 'exhortatives' and 'vocatives', are built upon imperative forms such as *(you) consider!*, while the adjective is the result of a predication (see below for details). The indicator of government (←) in the stative-verbant relation of *(you) consider!* is due to the imperative form; in fact, the stative *(you)* is usually omitted, hence the information salience stands onto the verb.

Examples (3-1) and (3-2) show two facts: first, not all verbants are verbs; second, within a single phrase the phrasal governor will always be a verbant.

Unergatives and unaccusatives

Valency is a great tool to analyse the stative-verbant relation, but it is not enough: we need a way to determine the information prominence— i.e., if the relation is generic (↔), a government (←) or a dependency (→). In this respect, we found the unergative hypothesis, originally posed by Perlmutter (1978), very useful. In fact, this hypothesis—fully presented in Perlmutter and Postal (1984)—gives account of both nominative-accusative and ergative-absolutive languages. This is a crucial fact, for every formalism that aims to be valid in general, i.e. for both nominative-accusative and ergative-absolutive constructions, as invoked by Dixon (1994) in the appendix. The adpositional paradigm aims to be an answer to this need. In constructive terms, the unergative-unaccusative hypothesis states

Table 3.2: The unergative-unaccusative hypothesis

name	*relation*	*test*	*first actant*	*monovalency*
unergativeness	dependency	X does I	$\overrightarrow{X}$	$\overrightarrow{I}$
unaccusativeness	government	I happens to X	$\overleftarrow{X}$	$\overleftarrow{I}$

that intransitives does not exist, or, more precisely, the expression 'intransitive verb' is a hyperonym for either an 'unergative verb' or an 'unaccusative verb'.

When a single stative-verbant relation is analysed, a simple question should be posed, counting as an informal test of ergativeness: does the stative *do* the verbant or does the verbant make the stative *happen*?

If the stative *does*, then the relation will be of dependency (→), as the information prominence stands in the stative: in the terms introduced by Perlmutter (1978), the actant will be *unergative* ($\overrightarrow{X}$).

Vice versa, if the stative *happens* to the verbant, the relation will be of government (←), as the information prominence stays in the verb, and the actant will be *unaccusative* ($\overleftarrow{X}$). The correspondent monovalent verbant construction will be unergative ($\overrightarrow{I}$) and unaccusative ($\overleftarrow{I}$), respectively.

It is worth noticing that Perlmutter and Postal (1984) referred exclusively to monovalent verbant constructions, while we adopt the terms in general, but always referred to the *first* actant, the first stative-verbant relation in terms of valency value. Table 3.2 summarises the hypothesis and explains the symbols used for the first-valent actant and the monovalent verb.

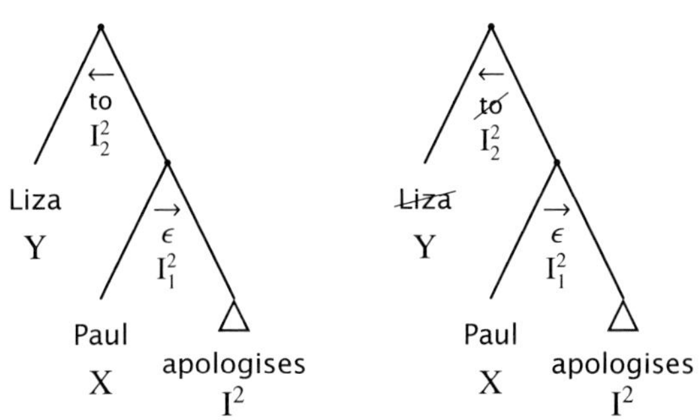

Figure 3.7: The trees of *Paul apologises (to Liza)*

(3-3a.) Paul apologises to Liza.

(3-3b.) Paul apologises.

Figure 3.7 shows the adtrees of (3-3ab). Paul *does* the apologies, hence the dependency ($\rightarrow$), while the apologies *happen* to Liza, who simply receives them, even if she does not want to; therefore, her stative-verbant relation will be of government ($\leftarrow$). In (3-3b), the verb *apologises* is used as a monovalent one: if the first-valent actant is unergative ($\overrightarrow{\mathrm{X}}$) the correspondent monovalent verb will also be unergative ($\overrightarrow{\mathrm{I}}$). Note that they are all character raisings.

From the formal point of view, the reader is invited to notice how the valence construction operates: since 'apologise' is bivalent (I^2), the valence construction is an arrow from the adtree 'apologises' to the adtree depicted on the left of Figure 3.7. This arrow comes from the family of arrows given by the valence construction: the family is indexed by ordered pairs of adtrees, the first component being 'Paul' and the second component being 'Liza'. The valence construction, in this case, denotes the first component by the empty adposition, ϵ, while it denotes the second component by the 'to' adposition. The grammar characters of these adpositions are I_1^2 and I_2^2, meaning that the whole construction takes two indexes and it operates starting from a morpheme of class I^2—the specific way to denote indexes depends on the source of the family of arrows, i.e., 'apologises'.

Example (3-3b) is the result of a transformation which hides the second index: in fact, no valid adtree exists when the second index is absent, as the valence construction would be undefined—it requires two indexes and only one is available, so the construction is non well-formed. By the way, this is the reason why the valence construction has its graphical shape: we need to put adpositions inside the tree, and adpositions may be non-empty; furthermore, some branches may disappear because of hiding. Then, it is convenient to adopt the binary tree format, which allows to put in evidence the single components and, eventually, to hide the pieces which are subsumed in the textual representation.

(3-4a.) Paul broke the vase.

(3-4b.) The vase broke.

(3-4c.) Paul bought a new vase.

Examples (3-4ab) show the complementary case of (3-3ab). The adtrees of (3-4ab) are represented in Figure 3.8. In English, the verb *broke* can be used either bivalently and monovalently. If monovalent, the cancelled actant will be

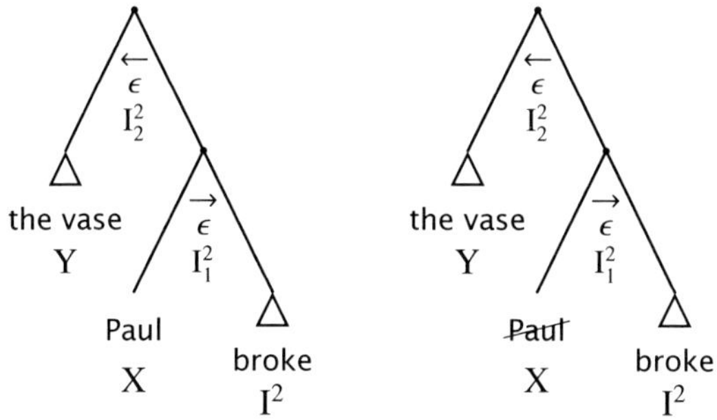

Figure 3.8: The trees of *Paul broke the vase / The vase broke*

the active one (3-4b), then the relation will be of government (→). Therefore, both the actant ($\overleftarrow{X}$) and the verb ($\overleftarrow{I}$) will be unaccusative. Again, all grammar characters have been raised.

Fillmore (2003a) reminds us that traditional grammarians of German have precise terms to distinguish bivalent phrases like (3-4a) from phrases like (3-4c): the classes of *verba affecta* (like 3-4a) and *verba effecta* (like 3-4c):

> One example of a 'covert' grammatical distinction is the one to which traditional grammarians have attached the labels '*affectum*' and '*effectum*', in German '*affiziertes Objekt*' and '*effiziertes Objekt*'. The distinction, which is reportedly made overt in some languages, can be seen in sentences 1 and 2: (1) John ruined the table; (2) John built the table. (Fillmore, 2003a, 27)

The adpositional paradigm gives a clear and precise account both of covert and overt grammatical distinctions, as the one referred by Fillmore in his classic article just quoted.

We can understand the terms 'unergative' and 'unaccusative' if we adopt a typological perspective: if a language normally adopts a nominative-accusative strategy, the morphologically marked case will be the accusative; by contrast, if a language adopts the ergative-absolutive strategy, the morphologically marked case will be the ergative (Dixon, 1994, 56–69). In constructive linguistic terms, in the first strategy the marked case will be the second valency (accusative), while the marked case will be the first one (ergative) in the second strategy. [10]

(3-5a.)	Ngarrkun	-tu	ari	kanyirr.
	wallaroo	-ERG	*eats*	*grass*

(3-5b.)	Ngarrkun	ari	-li	kanyirr	-ku.
	wallaroo	*eats*	-VALMINUS	*grass*	DAT

In Kalkatungu, an Australian language, the ergative-absolutive strategy is quite clear (Blake, 1990, 43). In (3-5a), the wallaroo (a kind of large kangaroo) is an agent, willing to eat grass, and the correspondent construction marks the first actant with an ergative morpheme. In (3-5b), the wallaroo is a patient, as the eating of grass happens to him: the verbant morpheme *-li* is used to decrease the value of valency (VALMINUS). [11]

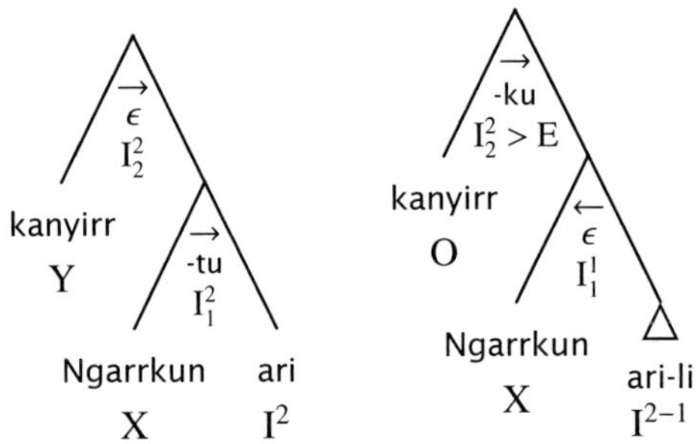

Figure 3.9: The trees of *Ngarrkun-tu ari(-li) kanyirr(-ku).*

Figure 3.9 shows the trees of (3-5ab). Consider that the morphology of words was hidden (△) as it is not the current focus here (see Section 5.1, devoted to transference); similarly, the notation $I_2^2 > E$ is also a sign of hidden information (see Section 4.1, devoted to epsilon-transformation). Instead, the difference among stative-verbant indicators should be noted here, because it signs unergativeness (3-5a) and unaccusativeness (3-5b).

Something similar happens in Japanese—a language typologically distant from Kalkatungu (Yamasaki, 2000, 27).

(3-5c.)	Inú	ga	hasítte	imásu.
	dog	AGENT	*runs*	DURATIVE

(3-5d.)	Inú	wa	hasítte	imásu.
	dog	TOPIC	*runs*	DURATIVE

This minimal pair shows a neutral form (3-5c), with the use of postposition *ga*, and the marked form (3-5d), with the use of postposition *wa*. In English, (3-5c) can be translated as *a dog is running* while (3-5d) can be something like *that dog is running* or *the dog, that we already talked about, is running.*

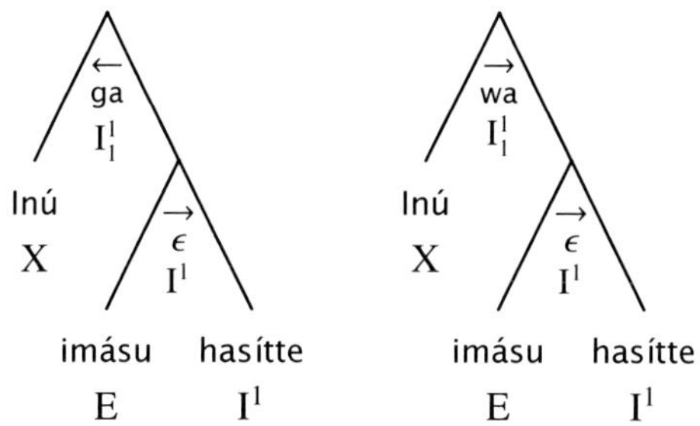

Figure 3.10: The trees of *Inú {ga|wa} hasítte imasu.*

Figure 3.10 shows the correspondent adtrees, which are very similar to the adtrees of Kalkatungu in Figure 3.9. So, in Japanese there are two specific postpositions to indicate the trajectory of information prominence: *wa* (→) and *ga* (←). These postpositions are very powerful, as they can verbify directly adjectives, as in the following pair of examples (Yamasaki, 2000, 60-61):

(3-5e.)	Yukí	wa	sirói	desu.
	snow	TOPIC	*white*	POLITE

(3-5f.)	Sonó	bàra	wa	akái	desu.
	that	*rose*	TOPIC	*red*	POLITE

The Japanese words *sirói* and *akái* are adjective by selection (see Section 5.2 for details; they are listed as *i*-adjectives in Japanese dictionaries for phonological reasons, giving a particular class of constructions as a direct consequence). It is worth noticing that this is a case of character imposition: in (3-5ef) the adposition *wa* verbifies the adjective, activating valency at the same time (Figure 3.11). Depicting the scene as (un)accusative or (un)ergative is strictly language-driven, as it depends on the culture where the language belongs.

Example (3-5) has shown that natural languages have always morphosyntactic strategies to change the scene according to the information prominence desired by the speaker.

Example (3-6) is another one of how typologically different languages map the information salience of the same scene in opposite ways.

(3-6a.) I sweat. (English)

(3-6b.) Sa-laksha. (Choctaw)

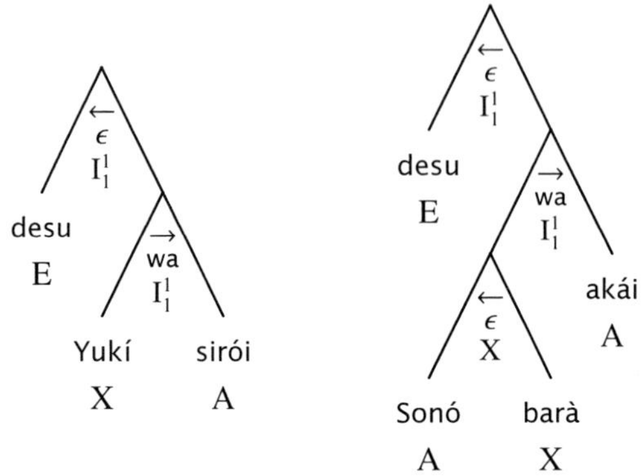

Figure 3.11: The trees of *wa*-based Japanese phrases with *i*-adjectives (3-5ef).

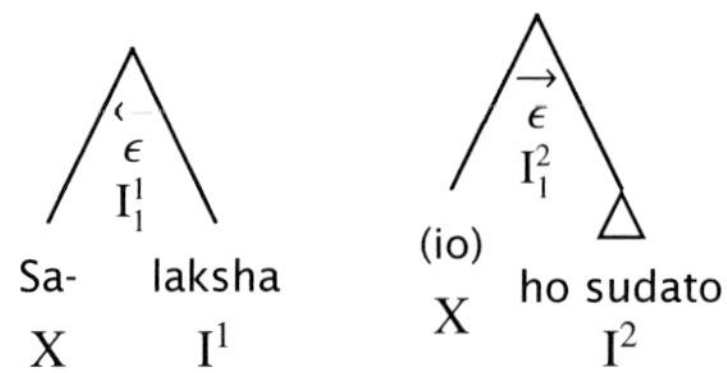

Figure 3.12: Trees of *Sa-laksha* and *(io) ho sudato*

(3-6c.) (io) ho sudato. (Italian)

Examples (3-6abc) show that even a simple event like sweating can be represented differently among different languages, as shown in Figure 3.12.[12] In Choctaw (a Native American language) the actant brings the objective marker of the first person *-sa-*, which indicates that the construction is unaccusative ($\overleftarrow{X}$), while in Italian the auxiliary form *ho* (from *avere*, 'to have') indicates that the construction is bivalent, with a stative-verbant dependency at the first actant ($\overrightarrow{X}$).

Sometimes the stative-verbant relation is left underspecified. For example, in Italian there are some verbs that accept both auxiliary constructions, with the verb *avere* (unergativeness: $\overrightarrow{I}$) or with the verb *essere* (unaccusativeness: $\overleftarrow{I}$).

(3-7a.)	Paolo	ha corso	per un'ora.
	Paul	$\overrightarrow{\textit{run}}$	*one hour long*

(3-7b.)	Paolo	è corso	a casa.
	Paul	$\overleftarrow{run}$	*home*

(3-7c.)	Paolo	corse.
	Paul	$\overleftrightarrow{run}$

In (3-7a) the underlying scene depicts Paul as willing to have a run for one hour, while in (3-7b) the scene is that Paul has run home abruptly (e.g., for some emergency). What about (3-7c)? As there is no auxiliary construction at work, the information prominence is left underspecified.

We will deal with the auxiliary verbant constructions in detail in the next chapter. However, the reader is invited to note that within the adpositional paradigm there is no distinct category of 'auxiliary verbs', rather stative-verbant constructions are used to express auxiliarity.

Grammaticalization and ontogenesis

The stative-verbant relations are the most fundamental ones in all languages, and for this reason they are the first to be learnt by children. In fact, we can consider constructions from a diachronic point of view as "historical relics…a salvaged part of some larger construction" (Tomasello, 2003, 104), but we can also consider constructions from an ontogenetic perspective. It is interesting to see that, as constructions are rote-learnt and lexically bounded, what is produced in the early stage of children's speech is exactly the most prominent information.

For example, the so-called 'word combinations' (18 months) are construction where the only groups produced are often the stative and the verbant, as shown in Figure 3.13:

(3-8a.) ball table.

(3-8b.) (the) ball (is on the) table.

It is worth noticing that the circumstantial character of *table*, imposed by the preposition *on*, is not actually used in the children's construction (Figure 3.13). Finally, the 'word combination' structure—having two distinct elements with a relevant collocation—can be reduced to a single structure where the only morphemes at work are the lexemes, i.e., the ones with the most relevant semantic load (Figure 3.14). [13]

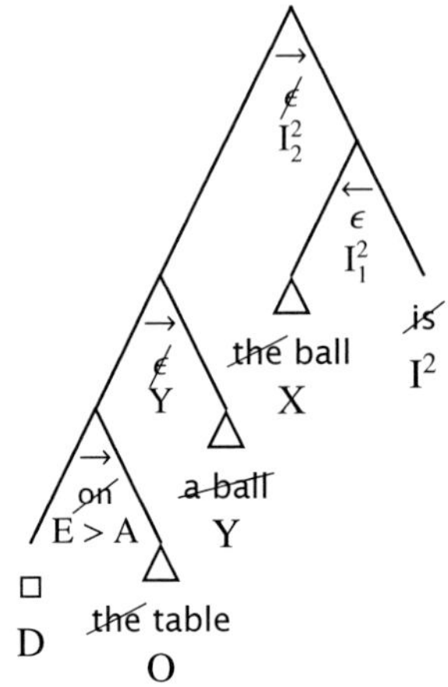

Figure 3.13: The underlying tree of *(the) ball (is on the) table*

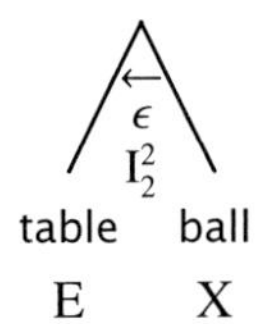

Figure 3.14: The final tree of children's speech *ball table*

Afterwards, starting from 18 months, children show more flexible constructions, called "pivot schemata" (Tomasello, 2003), such as the following ones:

(3-9a.) more milk.

(3-10a.) more grapes.

(3-11a.) ancora mela (*more apple*).

(3-12a.) ancora pasta (*more pasta*). [14]

It is quite obvious that the underlying structure is something like this:

(3-9b.) (I want) more milk.

(3-10b.) (I want) more grapes.

(3-11b.) (io voglio) ancora mela (*(I want) more apple*).

(3-12b.) (io voglio) ancora pasta (*(I want) more pasta*).

In this respect, English and Italian seem to be very similar. Figure 3.15 shows the correspondent adpositional trees, where the final one (right) is very similar to a prototypical unaccusative form ($\overleftarrow{I}$).

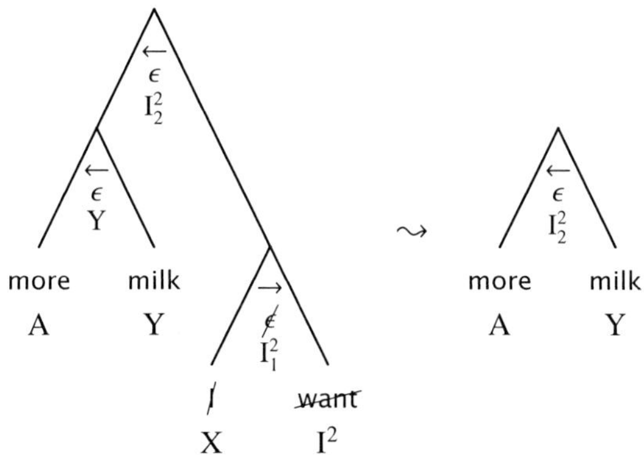

Figure 3.15: The trees of *(I want) more milk*

(3-13a.) juice gone.

(3-13b.) (the) juice (is) gone.

In a constructive linguistic perspective, we can see how much a 'word combination' is similar to a so-called "intransitive construction" (Tomasello, 2003, sic!), such as (3-13ab), because of the similarities of the respective adtrees (compare Figures 3.13 and 3.16).

It is also clear that information prominence is acquired very soon by children. For instance, in Tofi (a Samoan language) the following minimal pair is attested (Tomasello, 2003):

(3-14a.)	o nofo	pepe.
	sits	*doll*

(3-14b.)	o nofo	ia	pepe.
	sits	EMPH	*doll*

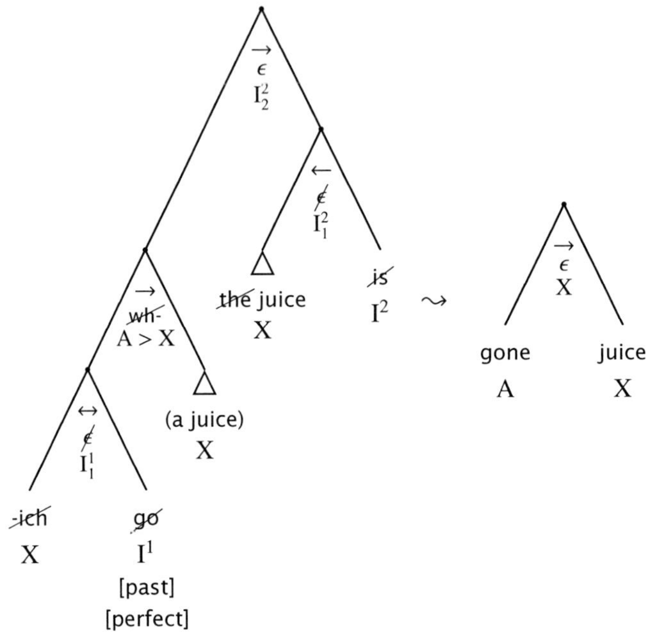

Figure 3.16: The trees of *(the) juice (is) gone*

Examples (3-14ab) mean something like *(my) doll sits*. It is interesting that Samoan children have an adposition (*ia*) exactly to reverse the trajectory of information prominence from government (←, without morphological marking) to dependency (→, with a special morphological marker). Figure 3.17 shows that the correspondent trees are quite simple in our perspective.

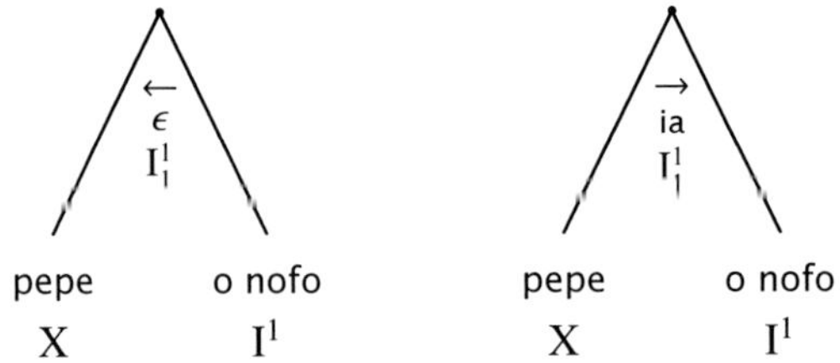

Figure 3.17: Trees of *o nofo (ia) pepe*

Compared with grammaticalization, the constructional arrow in ontogenetic acquisition is simply reversed. In fact, grammaticalization transforms lexical expressions (elastic groups of morphemes, e.g., *in the side of*) into fixed expressions (groups of morphemes to be treated as a single adtree leaf, e.g., *in side*) and finally to adpositions (e.g., *inside*): this is a process that starts from constructional morphology (adposition-based) to arrive to constructional syntax (collocation-based, i.e., with adpositions as epsilons). Conversely, ontogenetic acquisition shows constructions that start from syntax (statives and verbants, no modifiers, no connectives) in order to arrive at morphology (fully morphemic elasticity) during the adult stage of life.

A complete treatment of children's speech in constructive linguistic terms would require (at least) a whole book devoted only to this interesting topic. What we wanted to show here is that the adpositional paradigm can clarify some structures otherwise puzzling.

3.3 The adjunctive-stative relations

Adjunctives are stative modifiers. In a broad sense, the adjunctive-stative relation is a grammaticalization of the stative-verbant relation. However, before to deal with this grammaticalization of the correspondent adpositional trees—more rich and complex than what seems at a first glance—it is better to introduce the adjunctive-stative relations through simpler cases.

Attributive and predicative adjectives

In many languages of the world a distinctive subclass of adjunctives is found: adjectives. Moreover, in some languages—such as German—different constructions are at work if adjectives are *attributive* or by contrast *predicative*. Within the adpositional paradigm, this dichotomy represents the different trajectories of the information prominence in the adjective-noun relation: if the adjective is *attributive*, the information prominence stands in the noun, and hence the relation will be of government (←); vice versa, if the adjective is *predicative*, the information prominence stands in the adjective, because there is a predication (i.e., a stative-verbant relation) hidden under the adjective itself—we will deal with hidden predications in the next chapters. Consequently, predicative adjectives will relate to nouns with dependencies (→).

Some Romance languages show this distinction clearer than other languages. For example, in Portuguese:

(3-15a.) fortes ventos (*strong winds*)

(3-15b.) ventos fortes (*(only the) winds (that are) strong*)

In Portuguese, the different collocation into the syntagmatic axis of (3-15ab) shows precisely the different trajectories of information prominence: in (3-15a) the adjective *strong* is attributive, while in (3-15b) the same adjective is predicative. Figure 3.18 shows on the left the attributive adjective (3-15a) and on the right the predicative adjective, which conceals a predication.

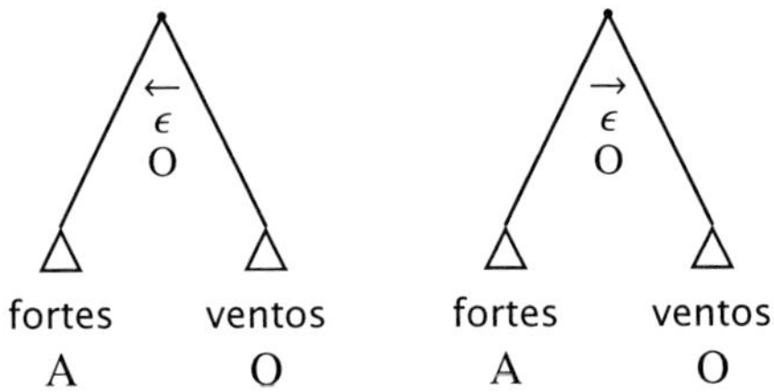

Figure 3.18: Minimal structural pair *fortes ventos* and *ventos fortes*

The difference in the adjective-noun collocation of Portuguese derives from two diverse Latin constructions. In Latin, the English group *our memory* can be rendered with the following constructions:

(3-16a.) nostra memoria (*somebody's memory we have*).

(3-16b.) memoria nostri (*the memory about us by somebody*).

It is worth noticing the different collocations: in (3-16a), the *attributive* adjective *nostra* is put **before** the noun *memoria*; in (3-16b), the *predicative* genitive *nostri* is put **after** the noun *memoria*. Portuguese has inherited the two collocations as historical relics of two different Latin constructions.

Analogously to the case of stative-verbant relations, sometimes also adjunctive stative relations are generic or underspecified:

(3-17a.) amor patris (*father's love (given to his children)*).

(3-17b.) amor patris (*(children's) love for (their) father*).

In Latin, *amor patris* can have two interpretations: the meaning of the parts is only one, but their senses as a whole are (at least) two. [15] In the first interpretation

(3-17a) the trajectory starts from the father (*patris*), which is the most prominent element because it gives love: therefore, the genitive-nominative relation (*patr-is* being the genitive, *amor* the nominative) will be a dependency (→).

By contrast, in the second interpretation (3-17b) the information prominence stands onto the love (the nominative *amor*), as the father receives love by somebody else—most probably, his children, him being a father. The important fact is that the relation is a government (←). [16] Figure 3.19 shows the two possible interpretations of *amor patris*. If it is not possible to determine the right interpretation, the information salience trajectory should be left underspecified through the use of the appropriate indicator (↔).

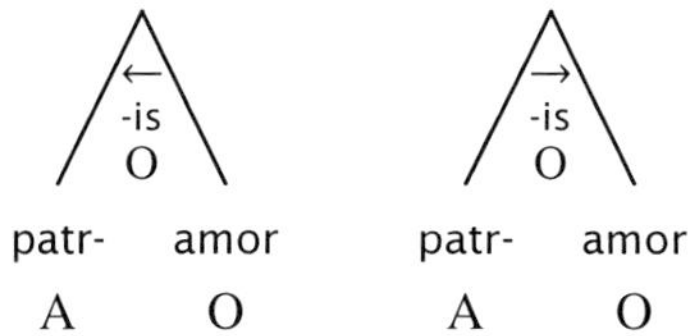

Figure 3.19: The two possible interpretations of *amor patris*

The dichotomy 'attributive *vs.* predicative adjectives' has a lot of interesting consequences: in particular, it will be reprised when dealing with correlative clauses (Section 4.6).

Other important adjunctives

Adjectives are not the only class of adjunctives used by languages. In fact, a lot of languages have adjunctive subclasses apart. [17] Here, with the word 'apart' we mean adjunctives that concur to the formation of construction in such a way not reducible to adjectives. Of course, adjunctive subclasses are language-specific: for instance, some languages have articles, some others not. It is impossible to treat them constructively into a single book—not to mention a section—even within a single language. The aim of this section is to explain how to build adpositional trees with English non-adjective constructions belonging to the adjunctive category. Therefore, what follows is highly partial: delving in this issue is left as a further work.

In English, there are at least the following specific adjunctive categories:

1. determiners (such as *a*, *the*, *that*);

2. quantifiers (such as *all*, *both*, *every*)[18];

3. numerals (such as *one*, *half*, *a million (of)*).

It should be obvious by now that all adjunctives are dependents (*dep*) as they modify stative, which are governors (*gov*). What it is addressed in this section is how to determine the final grammar character and the trajectory indicator of the adjunctive-stative adtree. A general cue for understanding the information prominence is the kind of reference brought by the adjunctive itself.

(3-18a.) the elephant.

(3-18b.) that elephant.

In English, saying *the elephant* and *that elephant* is quite different: the definite article *the* refers to something known (even elephants in general), while *that* marks a stronger reference (a physical elephant that is visible by the receiver, for example in a safari, or even an elephant supposed to be known by the receiver, according to the speaker's beliefs).[19] Figure 3.20 shows the correspondent adtrees.

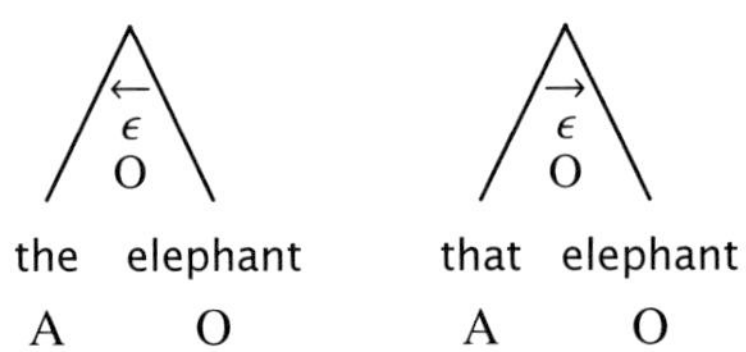

Figure 3.20: Trees of *the elephant* and *that elephant*

A stronger test comes from grammaticalization: if the adjunctive can be transformed into a stative with the omission of the original stative, that adjunctive will be in a relation of dependency (→):

(3-18c.) (I've seen) the one.

(3-18d.) *(I've seen) the.

(3-18e.) (I've seen) that one.

(3-18f.) (I've seen) that.

Figure 3.21 shows the adpositional trees of (3-18ef). It is intuitive to consider the adtree of (3-18f) as a single leaf—through a formal operation, called epsilon-transformation (see Section 4.1 in the next chapter for details). In (3-18ce) the numeral *one* is also used as a stative. It is important to notice, that this kind of rules are always language-driven.

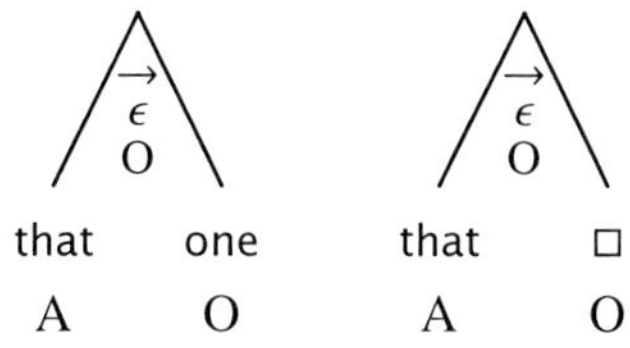

Figure 3.21: Trees of *that one* and *that*

Last, there is another kind of function that specific adjunctive categories exploits: the determination of grammar character in collocational morphemes.

For instance, the English morpheme *walk* is immediately recognised as a stative if a quantifier comes before, such as in the adjunctive-stative relation *every walk*. However, this kind of function—called transference—is so important that a section apart is devoted to it (see Section 5.1).

3.4 The circumstantial-verbant relations

Once stative-verbant relations are saturated, i.e., there are no more valency to be filled by actants into a phrase, and the adjunctive-stative relations are solved—by definitions to be solved within the actants—the circumstantials can be addressed.

Circumstantials are the modifiers of verbants: they are appended to verbant adtrees on the top being external arguments. The final grammar character, however, remains verbant—as it will be explained in the next chapter.

In English, adverbs are often used as circumstantials. Moreover, extra-valent stative groups (mostly head by a preposition) count as circumstantials. In Section 2.4 devoted to transformation, we have already encountered an example of a circumstantial:

(2-3g.) The janitor will open the door with this key for the Lady.

The group *for the Lady* cannot be advanced in the valency structure, henceforth it is extra-valent and so it is a circumstantial (Figure 3.22). There can be many circumstantials for the same phrase; in those cases, they are simply appended one over the other in the adtree.

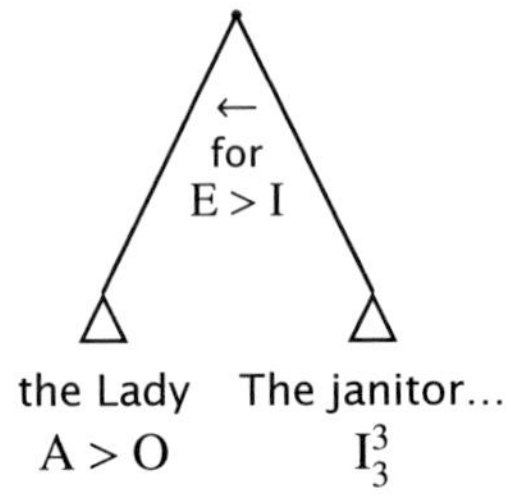

Figure 3.22: The circumstantial of (2-3g) in evidence

(2-3h.) The door will open quickly for the Lady without problem.

Figure 3.23 shows the adpositional tree of (2-3h). It is worth noticing that both the English prepositions *without* and *for* and the suffix *-ly* are considered adpositions that impose the grammar character within the adpositional paradigm. All circumstantial-verbant relations in (2-3gh) can be abstractely represented by the adtree in Figure 3.24.

In some languages, there are morphemes that are circumstantials by selection, i.e., without the need of any adposition: in those cases, what should be put is an epsilon (ϵ).

Examples (2-3gh) show circumstantial-verbant government (←): this is the default case, but it is not the only one.

For instance, some circumstantials show a considerable degree of freedom in collocation, i.e., they can be put in different places on the syntagmatic axis without clear, contrastive significance in terms of information prominence. Take for example the following minimal pair of Italian (not too distant from English):

(3-19a.) Domani vengo (*Tomorrow I'll come*).

(3-19b.) Vengo domani (*I'll come tomorrow*).

The two variants of (3-19ab) seem to be equivalent in terms of information prominence—although some informants may think that (3-19b) is marked, there is no widely accepted agreement about it. In doubtful cases such (3-19), there will be only one valid adpositional adtree (Figure 3.25).

All circumstantials seen until now are in a relation of government (←). However, in some cases the circumstantial is the most prominent information in the adtree, henceforth it is in a relation of dependency (→).

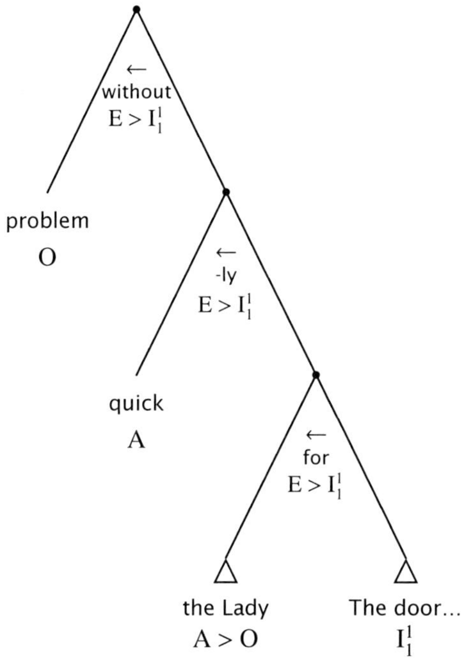

Figure 3.23: Many circumstantials at work in (2-3h)

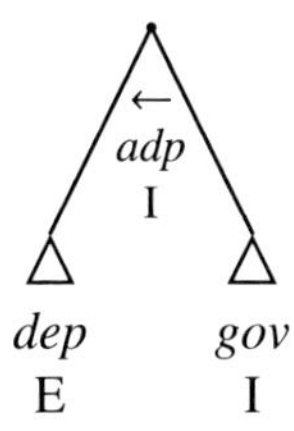

Figure 3.24: Abstract tree of circumstantial-verbant relations (2-3gh).

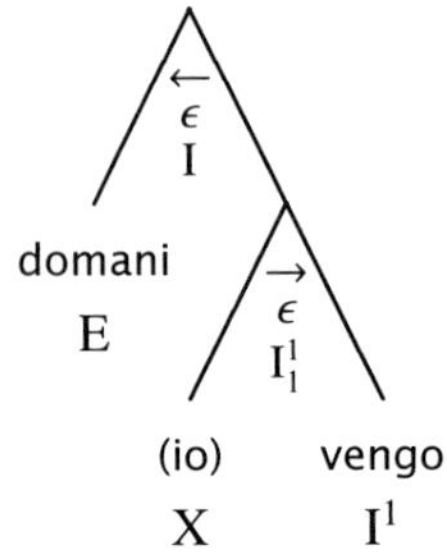

Figure 3.25: Tree of *Domani vengo* and *Vengo domani*

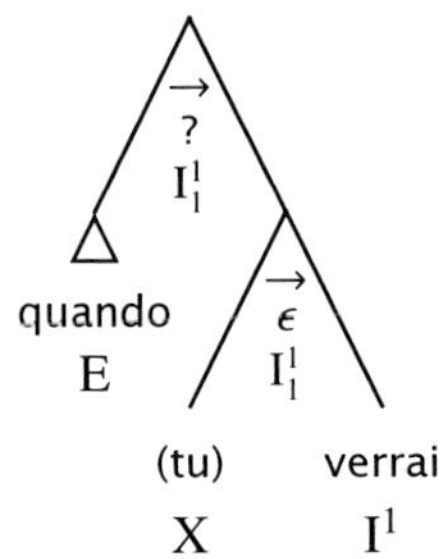

Figure 3.26: Tree of *Quando verrai?*

For instance, a circumstantial can be prominent if put into an appropriate construction, such as a question:

(3-19c.) Quando verrai? (*When will you come?*).

It is obvious that the most prominent information is in the circumstantial *quando* (when), as the circumstance of coming is exactly the information requested by the speaker (questions will be treated extensively in the next chapter). The trajectory of the circumstantial-verbant indicator will be chosen accordingly (→). It is interesting to notice how the adtree in Figure 3.26 is very similar to the one in Figure 3.25, referring to phrases (3-19ab), possible answers of (3-19c).

Finally, it is worth noticing that many languages mark the circumstance of temporality, causality, modality, location through ad-hoc circumstantials.[20]

In English, the words *tomorrow*, *today* and *tonight* are the result of the grammaticalization of Old English temporal construction *to* + O, where the stative has clearly a value of TIME, being *morrow*, *day* and *night* temporal nouns.

3.5 The circumstantial-adjunctive relations

In some languages—for instance English and French—circumstantial words are used to modify adjunctives. It is important to underline the fact that only circumstantial *words* can be used in such constructions, in these languages called 'adverbs': in other words, the inner adtree structure is of a circumstantial, while the outer structure shows the adverb as a adjunctive modifier, and the adjunctive, of course, will modify a stative.

(3-20.) that highly sensitive topic.

(3-21.) un vraiment beau paysage (*a really nice landscape*).

Examples (3-20) and (3-21) show stative groups with many adjunctives. In particular, there is a complex adjunctive group with a circumstantial word nested in. Figure 3.27 shows the adtrees of (3-20): on the left, there is the explicit adpositional tree, while on the right there is the adtree obtained after the epsilon-reduction. (3-20) also counts as an example of predicative adjective (*that*) and attributive adjective (the group *highly sensitive* taken as a whole).

The analysis of this attributive adjective group shows the circumstantial-adjunctive relation: the adjective *sensitive* is modified by the circumstantial *highly*. It is worth noticing that *highly* is a circumstantial as a *word*: his relation cannot be else than adjunctive (in fact, here there is no verbant, henceforth it cannot be a *phrasal* circumstantial in any case).

Figure 3.28 shows that the adtree of (3-21) is very similar to the one of (3-20). The main difference of the French example is the trajectory of information prominence: the phrasal indicators mark always a government (←), the most prominent element being the stative *paysage*.[21]

As a final note, please take into account the fact that special emphasis given by the speaker can reverse the indicators appropriately.

3.6 What is language structure?

Now it is possible to describe a language in constructive linguistic terms. In order to have a language, the following ingredients are needed:

1. the rules to build adtrees, in particular valency (syntax, explained in Chapter 2)—these rules are called *constructions* in our formal model;

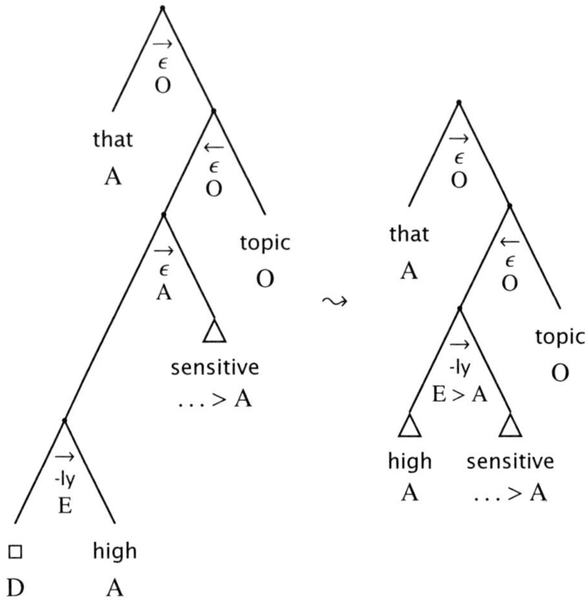

Figure 3.27: Trees of *that highly sensitive topic* (explicit and ϵ-transformed)

2. information prominence and trajectory indicators (morphosyntax, examined in Chapters 2 and 3);
3. grammar character (just explained here, in Chapter 3);
4. the rules to transform an adtree in another adtree (transformation; it was sketched in Section 2.4 and will fully explained in Chapter 4)

Transformations need a treatment apart, since they constructively describe complex constructions, i.e., constructions made of more phrases.[22] In other words, complex constructions are constructions that put together more verbant governments and their in-phrasal constructions (just presented in this chapter). The next Chapter 4 is devoted to this topic. Furthermore, in the next chapters we will deal with *transference*, a special case of transformation.

After that, we will be ready to construe the dictionary of a specific language, i.e., a structure with properties where the four ingredients of that particular language structure can be mapped into (see Chapter 5). In fact, in order to have a grammar we need a language structure and a dictionary that, together, generate

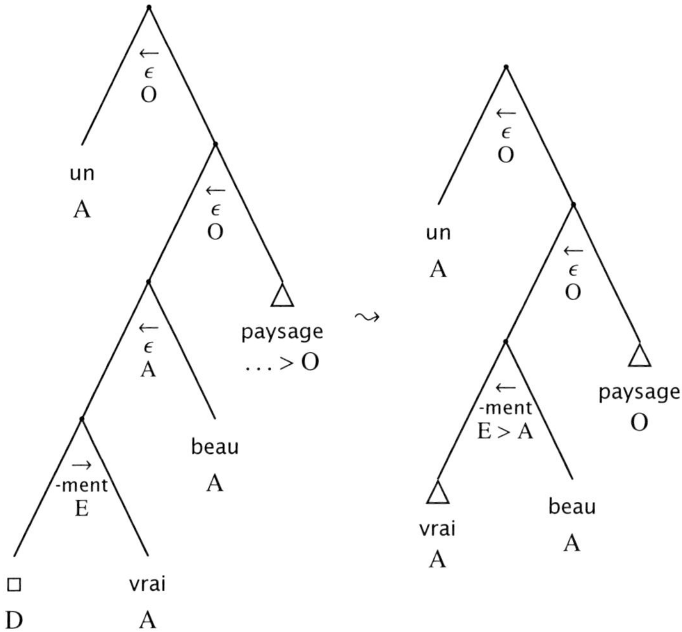

Figure 3.28: Trees of *un vraiment beau paysage* (explicit and ϵ-reduced)

the space where to instantiate the rules of the language itself and where to obtain the whole set of possible correct productions.

Chapter Four

Transformations

While syntax was presented in Chapter 2 in a very general and abstract way, in Chapter 3 we have instantiated it within linguistic constructions. Constructions arise in a given language as products of grammaticalization. In this chapter, we give account of grammaticalization in terms of adtree *transformations*, i.e. in terms of formal rules where a given adtree becomes another adtree. Generally, grammaticalization implies a *reduction* of the resulting adtree in comparison with the starting one—as we will see in detail, sometimes there are more than just two adtrees involved in the process.

Under a linguistic point of view, grammaticalization phenomena—and hence transformation patterns—are highly language-bound. Even if we compare typologically similar languages (say, French and Italian) transformations can be very different. [1] In particular, morphology (how morphemes combine in order to form words) is very specific within each language, mainly because of phonotactics (the allowed sequences of phonemes of that language) and graphemics (its writing system tradition). [2]

But grammaticalization phenomena do not pertain only to morphology, but also to syntax, in the sense of combination of words, expressions, phrases. We will see that adtree transformations give a precise and flexible account of the inter-phrasal level—or sentence level, if you prefer—of language analysis.

In principle, there is no difference between grammaticalization at morphologic and syntactic levels: the rules of transformations are the same. Nonetheless, it is useful to formally separate the purely linguistic level—where the rules that described constructions are the same, regardless of the languages—from the morphemic level, where the degree of grammaticalization is so high that it cannot be separated from the specific language in analysis, in particular for the constraints carried by the semantic and syntactic traits of the morphemes. However, in practice we play moves on the linguistic and morphemic levels in parallel, when dealing with transformations, for example when we have to represent correlative clauses.

Under a formal point of view, transformations are all endofunctors in the category of adtrees—see Appendix B.3 for the details. In practice, we distinguish three types of transformations, along with their purposes: tree transformations, redundancy transformations and epsilon transformations.

Tree transformations (or τ-transformations) are those transformations taking an adtree in input and giving a differently shaped adtree in output, therefore they operate at the linguistic level. When the chain of τ-transformations is over, redundancy transformations (or ρ-transformations) are in charge.

A ρ-transformation generally operates on the morphemic level, as it deals with the final morphemes —that is, proper morphemes, that cannot be expanded in adtrees any further. In other words, generally it takes the final τ-transformed adtree in input and gives in output an adtree with the same structure, with at least one different morpheme, for redundancy reasons. So, a ρ-transformation can give account of agreement in number, gender, case, or person, depending on the particular construction at work.

Finally, epsilon transformation (or: ϵ-transformations) are a useful way to hide empty adpositions (ϵ) and void governors and dependents ($\square$). As it will be clear in the next section, the treatment of empty adpositions and void leaves, both governors and dependents, is different since, despite their name, they are not transformations but just a convenient way to write adtrees.

4.1 The epsilon transformation

An epsilon transformation is not a kind of linguistic rule, rather it is a merely notational manner to write more compact adtrees, without losing relevant information. This implies that ϵ-transformations are not related with the other kinds of transformations, and hence their explanation needs only what already seen in the previous chapters. For this reason, we have chosen to explain them before the other two types of transformations.

As already seen, sometimes grammar characters are signed with a morphemic marker, sometimes not—in this last case they are purely collocational. In particular, a collocational strategy (which pertains to syntax, not to morphology) is more likely to be found in adpositions and dependents, rarely in governors: after all, it is obvious that the governor of all the morphosyntactic structure normally is elicited—however, an exception was offered in (3-18f).

A purely collocational adposition will be signed by an epsilon (ϵ), while an adtree leaf (either governor or dependent) by a box ($\square$). The difference is not only notational, but hence it respects a conceptual difference: while adpositions are hooks in the adtrees, governors and dependents are *themselves* adtrees, where the adtree coincides with a single element, exactly the governor or the dependent. [3] A way to hide epsilons and boxes when needed revealed to be very useful when we found the model. We want to stress the fact that ϵ-transformations are a useful notational artifact, but nothing more: no information is lost but only hidden, essentially for human readability. Hence, when we will talk about 'cutting off' or 'eliminate' what we really mean is cancelling from the actual adtree drawing.

The only way to eliminate void *dependents* is when their adposition is empty too (Figure 4.1). We have already seen an instance of it in the previous chapter in Figure 3.13, referring to example (3-8).

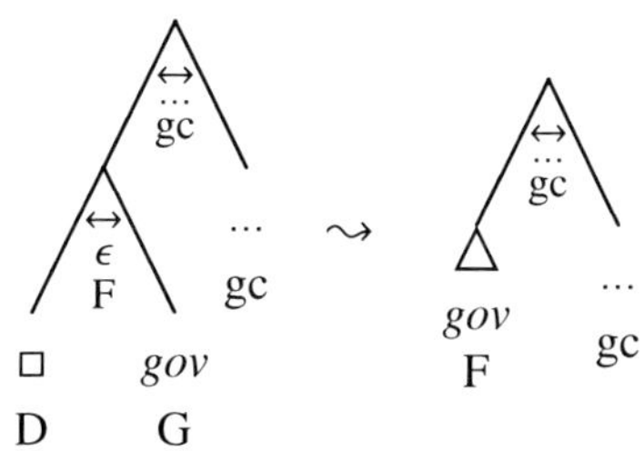

Figure 4.1: Abstract trees of epsilon transformation at the first level

In general, there are two possible cases, depending on the depth of the adtree to be transformed. Let *adp* be a morphemic adposition (i.e., with a concrete linguistic marker on the syntagmatic axis) and ϵ be a collocational adposition (i.e., no marker on the syntagmatic axis). The epsilon transformation at the first level (in brief: ϵ_1-transformation) is applied in all cases of grammar character imposition—recall Figure 3.1 at the beginning of Chapter 3. Figure 4.1 shows an ϵ_1-transformation: the non-transformed adtree is on the left, while the transformed tree is on the right. It is worth noticing that the transformed tree becomes a leaf; we have already seen an instantiation of this kind of transformation, about children's speech utterance *ball table* (3-8).

Sometimes we have performed a different kind of epsilon transformation, which comprises two levels of the adtree. For this reason, it is called epsilon transformation at the second level (in brief: ϵ_2-transformation). For instance, this is applied to the circumstantial-based relations in general, i.e., the circumstantial-verbant relations, e.g., (3-19), and the circumstantial-adjunctive relations—e.g., (3-20) and (3-21). Let G_x be the grammar character of both governors involved (gov_1 and gov_2). The abstract rule is shown in Figure 4.2. The ϵ_2-transformation

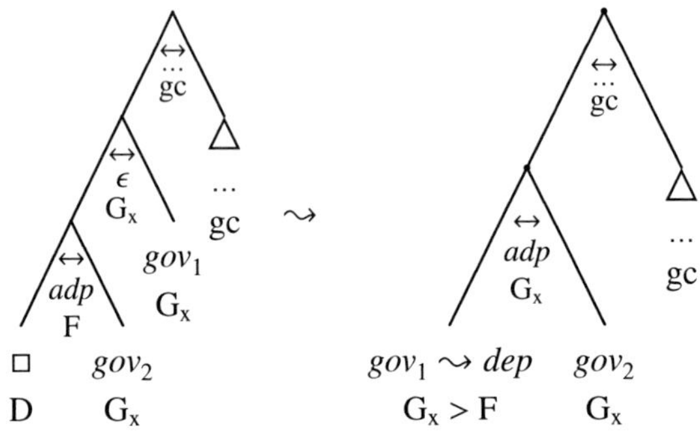

Figure 4.2: Abstract trees of epsilon transformation at the second level

shows what happens in case of 'clashing' between character imposition and character raising: at the first step, ϵ_1-transformation is applied (character imposition) while at the second step character raising 'wins', so that the adposition is put under the upper level (Figure 4.2).

There is no way to eliminate void *governors*, even if their adpositions are empty: their information is too important. However, we have found a compact notation for such cases.[4] Let us suppose to have a character raising with an void governor and an empty adposition (which is mandatory) but a non-void dependent (Figure 4.3). We can obtain a real-world case as such from (3-18), where *that elephant* becomes *that one* and finally *that*—see Figure 3.20 for a comparison. This is the only way to obtain a new governor from a dependent: the reader is invited to note that the final grammar character is the one of the *old* governor, not the one belonging to the transformed dependent.

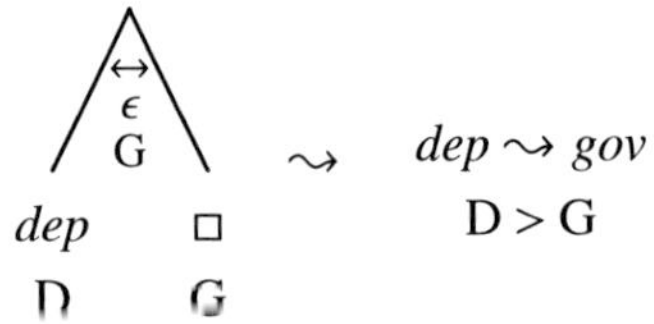

Figure 4.3: Abstract trees of how to compact a void governor

The symbol > is used in Figure 4.3 as a shortcut for an adtree whose only relevant information is the grammar character change of the only explicit morpheme at work in that particular adtree. In other words, the adposition is empty and either the governor or the dependent is void.[5]

Finally, it is worth noticing that in every epsilon transformation there is no assumption or modification on the trajectory of information prominence. [6] In case of compacted void governors, the indicator can be put over the final grammar character explicitly, if needed: e.g., D > $\overleftarrow{\text{G}}$.

4.2 The mechanics of tree transformations

Tree transformations are by far the most important type of transformations, as they give an account of grammaticalization phenomena, that can be described as adtrees transformed into adtrees. Formally, we use conjugate adtrees, where parts (leaves or larger subtrees) of the input adtree are permuted—for further details, see Section B.3. To explain how it works, let us recall example (2-3), we have previously seen. An obvious linguistic example of τ-transformation is the one that transforms a phrase from the active to the passive diathesis.

(2-3a.) The janitor will open the door with the key.

(2-3f.) The door will be opened by the janitor.

What we want to show is how τ-transformations starting from (2-3a) obtain (2-3f). It is a good example, as it has the advancement of the second actant (*the door*), the withdrawal of the first actant (*the janitor*), with the insertion of the specific preposition *by*, as well as the omission of the third actant (*with the key*). A further advantage is that this kind of active-passive construction is common in many language belonging to the Standard Average European Sprachbund.

Figure 4.4 shows the various steps. The reader is helped in distinguish conjugate adtrees from ordinary ones as the symbols in the conjugates are marked by a star (*). It is worth noticing that we are operating only at a quite abstract and linguistic level: in fact, every leaf is signed with a triangle (△). Within the class of every phrasal construction with that particular governor—i.e., morphologically well defined—we should start from the *saturated* construction, that is the construction with all actants in action.

The first step gives account of the omission of the third actant: please note that, although the resulting adtree *is* a linguistically valid adtree, it is a conjugate one, i.e., it has no sense out of the transformation chain, and therefore it is marked accordingly. The second step performs a *permutation* between the first and the second actant: their respective adpositions follow the leaves accordingly. The third step performs the *insertion* of the adposition requested by

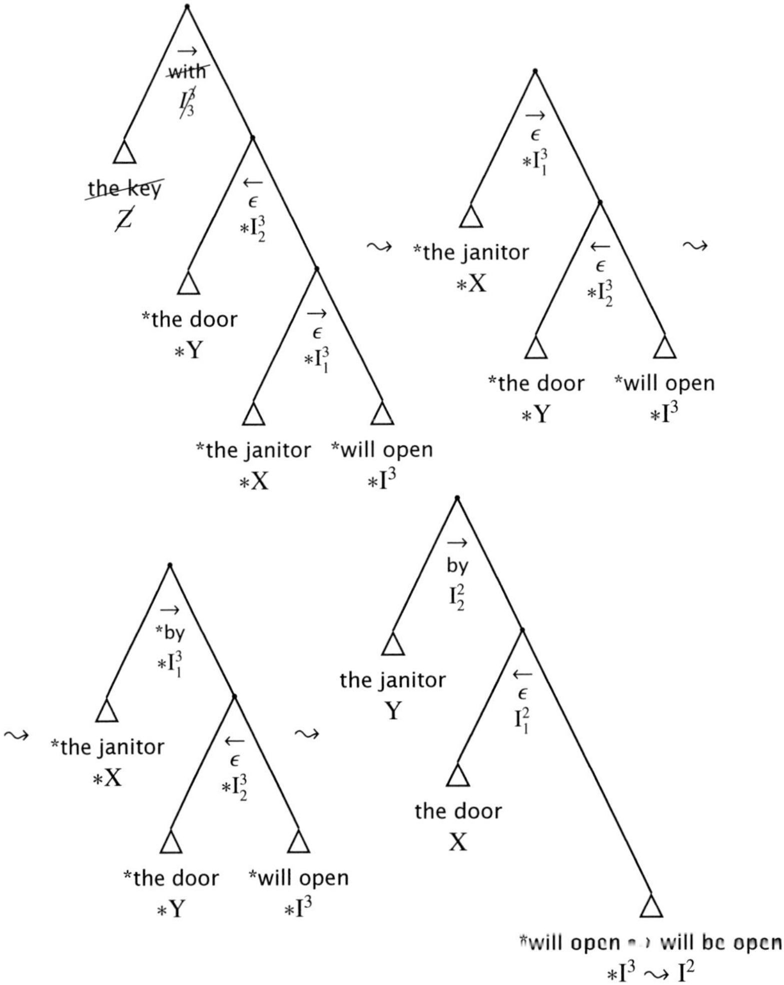

Figure 4.4: Example of tree transformation in an active-passive construction

the final passive construction, i.e., *by*. Finally, the fourth transformation operates at a morphemic level: it redefines the construction of the verbant form in order to satisfy the requested feature of being passive. The valency values are recalculated accordingly. The transformation from the active verbant *will open* to the passive one *will be opened* is explained later, as it involves the treatment of hidden predications.

4.3 How redundancy works

We should still give an account of agreement and, in general, redundancy, within the adpositional paradigm. In order to do so, we should redefine the notion of 'morpheme'. In fact, a morpheme is a double-side sign, analogous to phonemes: the first side is the signifier, i.e., the explicit mark on the syntagmatic axis will be called *morph*, while the other side is formed by every other trait, called *sememe*. Sememes usually convey semantic information, but also syntactic—such as case, semantic role, number, etc. For example, the morph of *walk* is 'walk', while MOVEMENT is one of its sememes. Typographically, the morpheme is written in italic, the morph in plain, while sememes are written in small capitals. Each morpheme has always one morph (a zero marked morph at the limit, signed through a box: □), and at least one sememe. Sememes are strictly language-specific, and they should be made explicit while writing the constructive dictionary—see Chapter 5 for details.

Redundancy is a linguistic phenomenon where a feature within a construction is marked more than the necessary, i.e., more than the minimum in order to convey the feature itself. For 'feature' we mean, very informally, every collection of sememes. Redundancy is a language universal: no natural language is free from redundancies. For instance, agreement is a very common type of redundancy, where the feature marked more than necessary is number, gender, case or person.

(4-1a.) Liza walk.

(4-1b.) Liza walks.

Example (4-1a) shows a phrase in English which is acceptable only under specific socio-linguistic conditions, exactly because it lacks the necessary redundancy, shown in (4-1b). Every good reference of the English grammar states that the verb of the third person singular of the simple present tense must end with *-s*, *-es* or *-ies* for phonotactic reasons (Alexander, 1988). Nevertheless, omitting the final *s* does not change in any way the meaning of the verbant or its linguistic structure, but it prevents to 'cross-check' the correctness of the expression.

Unlike the example of the previous section, here we are operating only at a morphemic level: from a purely linguistic level of analysis, (4-1ab) are the same. In general, we will write out only the sememes needed for our purpose in that moment. For example, we can rewrite (some of) the morphemes of (4-1) explicitly as follows:

(4-1c.) Liza [NOUN: PROPER, SINGULAR] walk-s [VERB: [TENSE: SIMPLE PRESENT: [3RD PERSON, SINGULAR]]].

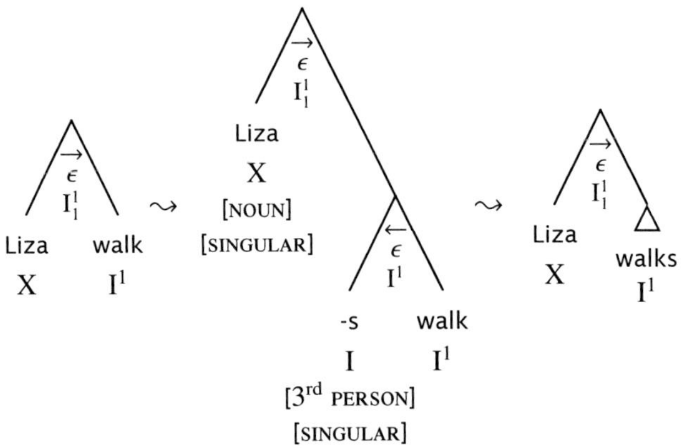

Figure 4.5: Redundancy of *Liza walks*

Example (4-1c) shows quite clearly that the explicit writing of *every* sememe pertaining a very simple phrase such as (4-1) is neither feasible nor useful, if possible at all.[7] It is the fact that within the construction of (4-1) *walk* is a VERB that can activate its subordinate sememe SIMPLE PRESENT describing tense, and then accordingly the rule of agreement of the third person singular between the verb itself and the (proper) NOUN *Liza*[8].

Figure 4.5 shows the correspondent adtrees: (4-1a) is on the left, while in the middle the agreement is solved by showing the matching sememes between the verbant (I) and the first actant (X). In fact, if the first actant is a NOUN: SINGULAR (it does not matter if PROPER or COMMON) and the verbant is a VERB: [TENSE: SIMPLE PRESENT], then a modifier with the sememes [3RD PERSON, SINGULAR] should be applied to the VERB, in this case -*s* (according to phonotactic rules written in the dictionary), in order to match the sememes NOUN: SINGULAR of the first actant. The trajectory of information prominence is that of government, as the morpheme *-s* does not change the grammar character of *walk*, which remains the most prominent information of this morphemic adtree.

An observation should be done at this point. It is not by chance that many redundancy phenomena are carved into each language morphology: after all, morphology is a clear sign of grammaticalization. This does not imply, however, that ρ-transformations cannot also occur at a syntactic level.

(4-2a.) Liza walks and Paul walks.

(4-2b.) Liza and Paul walk.

The construction headed by *and* as the main adposition turns its elements, i.e., governor and dependent, taken as whole, into a stative, as (4-2ab) show—the quotation in (4-2a) was put to help the reader in this respect.

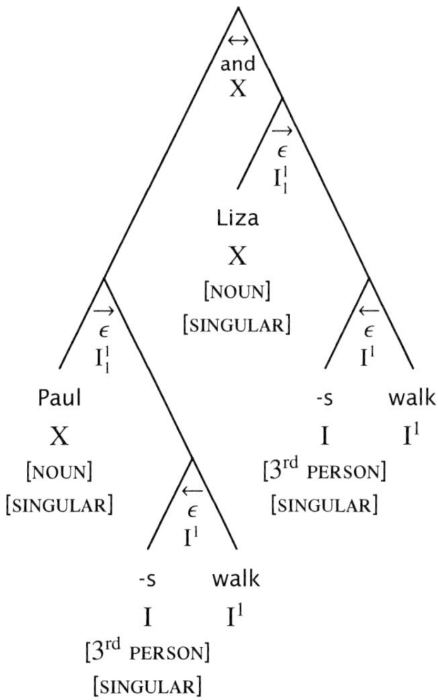

Figure 4.6: The adtree of *Liza walks and Paul walks*

Figure 4.6 shows the adpositional tree of (4-2a): there is a lot of redundant information in it, so it is far common to use the construction exemplified by (4-2b) instead. In this construction, where the governor and the dependent of *and* are both stative and SINGULAR, the resulting adtree will be PLURAL, as the agreement with the verb *walk* testifies, in this case through the *absence* of the morpheme *-s*—Figure 4.7.

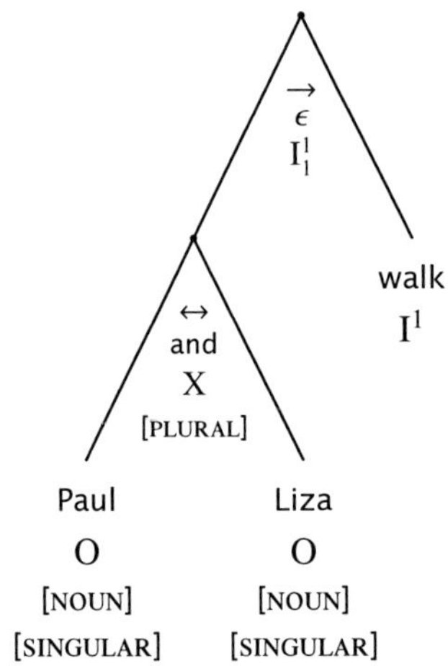

Figure 4.7: The adtree of *Liza and Paul walk*

In the sequel, we will show some of the most important and well-known phenomena that involve some types of transformations, such as passivization, normalisation, auxiliary constructions and so on.

4.4 To be and to have: that is the question

We show how to deal with auxiliary constructions that are common in languages belonging to the Standard Average European Sprachbund. In particular, in Romance languages the choice of the auxiliary verbant corresponding to the Latin *esse* and *habere* has a lot of important consequences. For this reason, they are presented together in this section. We will refer mainly to English for simplicity, but we claim that many observations can be easily extended to other language belonging to the same Sprachbund, after the obvious adaptations.[9]

To be, an unexpected bivalent verb

There are a lot of constructions based on *esse*, not only auxiliary in nature. We can list the following prototypical proper *esse*-constructions in English that show the patterns of use of *to be* as "a full verb when it combines with adjectives and nouns" (Alexander, 1988, 187). [10]

(4-3a.) Liza is a quiet girl.

(4-3b.) Liza is quiet.

(4-3c.) Be a quiet girl!

(4-3d.) Be quiet!

By contrast, in (4-4) auxiliary constructions are shown, that is *esse*-constructions where *to be* is the active verbant of the main phrasal construction, but at least one other non-active verbant is present. [11]

(4-4a.) Paul is studying maths.

(4-4b.) Paul is going to study maths.

(4-4c.) Maths is prepared by Paul.

(4-4d.) Maths is prepared.

(4-4e.) Paul is prepared.

(4-4f.) Be prepared!

We will see the constructions where *to be* is a proper verb at first. The *esse*-constructions puzzled Western philosophers since at least Aristotle (Moro, 2010). In our perspective, the main problem with these constructions, in particular where *to be* act as the so-called 'copula', is that *esse*-constructions have always been wrongly treated as monovalent constructions, while they are bivalent.

The adpositional tree of (4-3a) shows what should be evident: there are two stative involved in *esse*-constructions, not just one—Figure 4.8, left. The anomaly is in the information prominence: both the first and the second actants are not prominent, hence they are in a relation of government (←), while the prominence is in the adjective *quiet*, which is in a relation of dependency. [12]

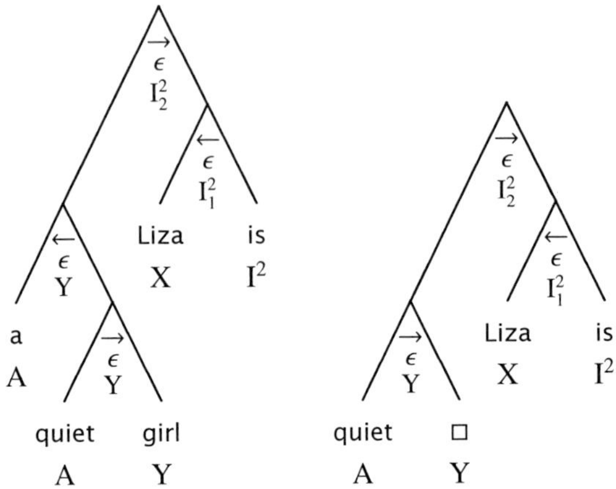

Figure 4.8: The adtrees of *Liza is (a) quiet (girl)*

This is confirmed by (4-3b): the information belonging to the second actant and not prominent is completely lost, and only a trace is conserved—signed by the box: □.[13] The reader is invited to notice that the information prominence of the adjective (→) is retained. Of course, we can obtain (4-3b) from (4-3a) with a series of tree transformations: this implies that the *esse*-constructions with adjectives, like the one in example (4-3b), are reduced forms of the main *esse*-constructions exemplified by (4-3a).

The imperative constructions exemplified by (4-3cd) are the product of further transformations performed on the first actant, which retains only the sememes of the SECOND PERSON: SINGULAR, with the morph unexpressed (or a zero morph, if you prefer). The corresponding adtrees are shown in Figure 4.9.

Now we address the auxiliary *esse*-constructions in English. What is an auxiliary construction? Within the adpositional paradigm, an auxiliary construction is a complex construction where the *auxiliary* (helping) phrase is governed by the active form of the verb *to be* while the second actant is the *meaningful* (helped) phrase: a transformed, or, more precisely, reduced phrasal construction where its predication is partially hidden—we say 'partially' because the verbant morph is still expressed, as in *studying* (4-4a) or *prepared* (4-4c).[14]

Figure 4.10 shows the adpositional tree of (4-4a), which exemplifies the present progressive construction in English. In particular, *is* is the morpheme of the auxiliary verb (symbol: I^2_{aux}) and *study* is the morpheme of the morpheme of the helped verb (symbol: I^2_{mean}). Please note that the adtree of the active and sat-

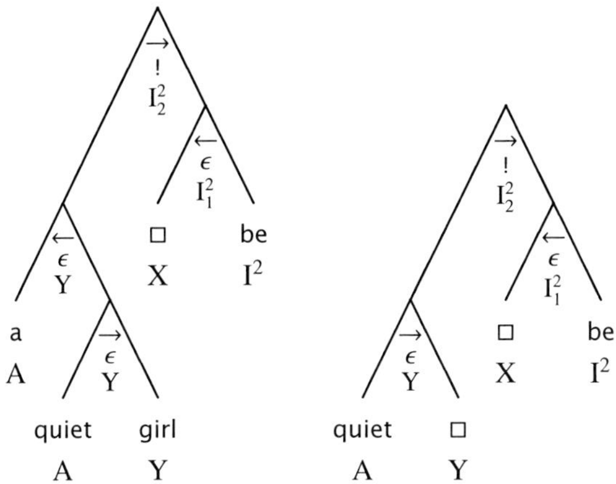

Figure 4.9: The adtrees of *Be (a) quiet (girl)!*

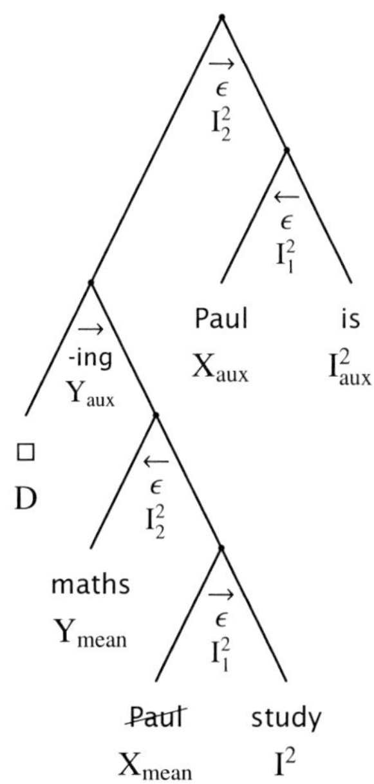

Figure 4.10: The adtree of *Paul is studying maths*

urated construction *Paul studies maths* is preserved—except for the redundancy verbal marker *-s*. In order to function properly, the English grammar requires that the verbant *studies* is de-activated by the adposition *-ing* by imposition of the grammar character. The second actant of the helped construction (symbol: Y_{mean}) saturates the second and last valency.[15] Finally, the morpheme *maths* is the result of a common form of grammaticalization called *lexicalizations*: in fact, even if morphologically it is a plural (*math-s*), it is used as a singular item, and hence it can be listed as a unanalysable lexical entry in the English constructive dictionary–see Chapter 5.

In the case of unaccusative verbs ($\overleftarrow{I}$), like *Paul is dying*, the second morpheme becomes void (symbol □).

The rendering of the *going to* future construction exemplified in (4-4b) is slightly more complex, as there are two auxiliary verbants, the first being an *esse*-construction while the second one is the proper *going to* construction (I^2_{aux+}), where the meaningful verb acts as the second actant—see Figure 4.11, left. This second auxiliary construction *going to* upon frequency of usage was completely grammaticalized, so that we can treat *-ing to* as a single adposition—see Figure 4.11, right—thus, the final adtree is drastically simplified. This grammaticalization is confirmed by the existence of the informal morph *gonna*, which has exactly the same function of *go-ing to* within this class of constructions.[16]

To confirm the double auxiliary construction, consider the phrase *I am looking at eat/eating an apple*: in this case, *looking* introduces its second actant via the *to* adposition, while the second actant is either *eat an apple* (governed by an infinitive) or *eating an apple* (governed by a gerund). There is a nuance in meaning if the infinitive is chosen instead of the gerund: if the actant is an infinitive, the action is indefinite in time; if the actant is a gerund, the action is in an (hypothetical) immediate future and the attention is focused on the process (duration) of the action. The general form of infinitive and gerund constructions are described later in this chapter.

Now we can address the passive construction in detail. The reader is invited to recall (4-4cd):

(4-4c.) Maths is prepared by Paul.

(4-4d.) Maths is prepared.

The corresponding active construction is *Paul prepares maths*. Let us call *Paul*, being the first actant of the active form which does the preparation, $\overrightarrow{X_{act}}$,

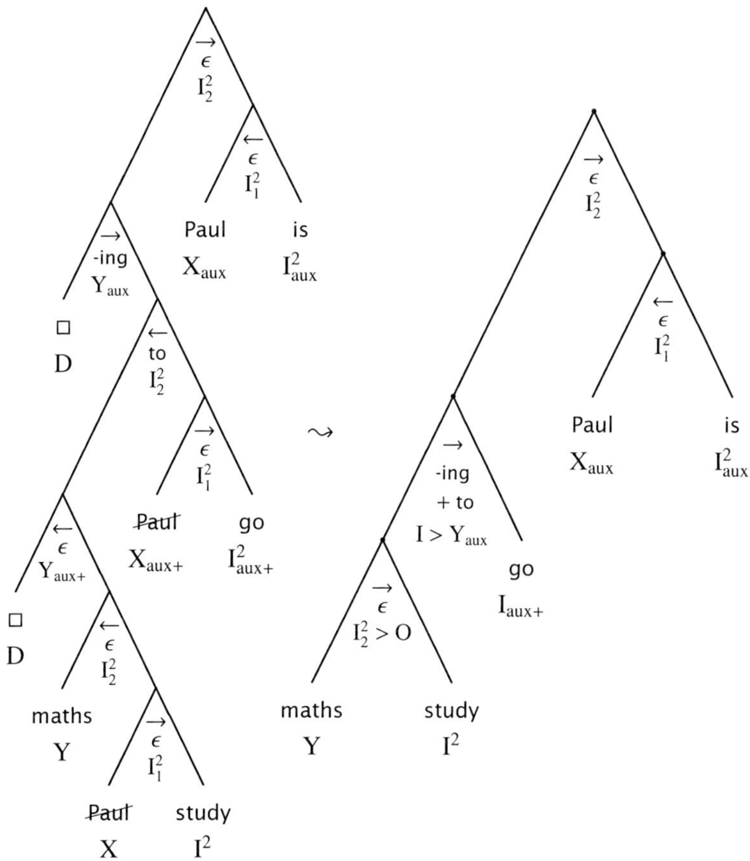

Figure 4.11: The adtrees of *Paul is going to study maths*

while *maths*, being the second actant of the prepared subject happened to be studied, $\overleftarrow{\text{Y}_{\text{act}}}$. The passive construction can be obtained by a chain of transformations. From them, we retain these abstract names for the two actants even in the passive form, in order to understand better how the passive construction works.

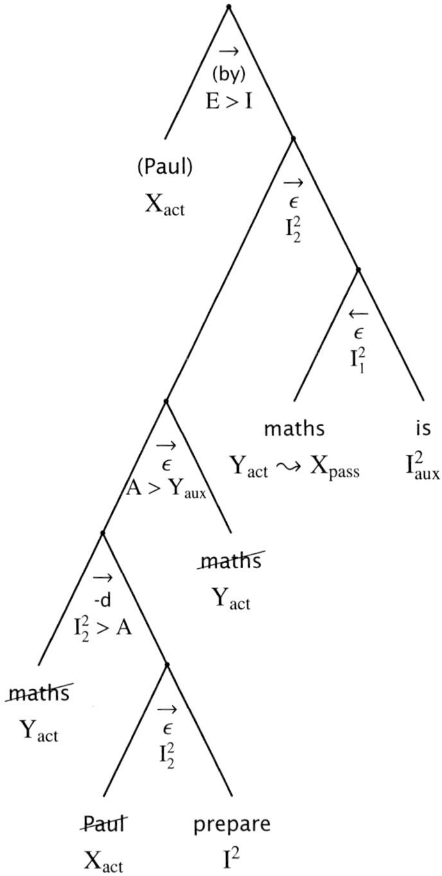

Figure 4.12: The adtree of *Maths is prepared (by Paul)*

Figure 4.12 shows the adpositional trees of the passive constructions in (4-4cd). The first obvious fact is the permutation of the two actants: in the passive construction *maths* passes from the role of the second to the first actant ($\overleftarrow{\text{Y}_{\text{act}}} \rightsquigarrow \overleftarrow{\text{X}_{\text{pass}}}$): please note that the trajectory of the information prominence does not change (→). Conversely, the first actant ($\overrightarrow{X_{act}}$) goes out of valency, becoming a circumstantial (E). This happens because the PREPARER *Paul* can be safely omitted, as in (4-4d).

(4-4e.) Paul is prepared.

(4-4f.) Be prepared!

It is interesting to see that, for purely semantic reasons, the first actant of the advanced form ($\overrightarrow{X_{act}}$) can be advanced in the passive form (4-4e), roughly meaning that Paul has *really* prepared the exam of maths or whatever: the second actant (Y_{act}) cannot be retrieved anymore. [17]

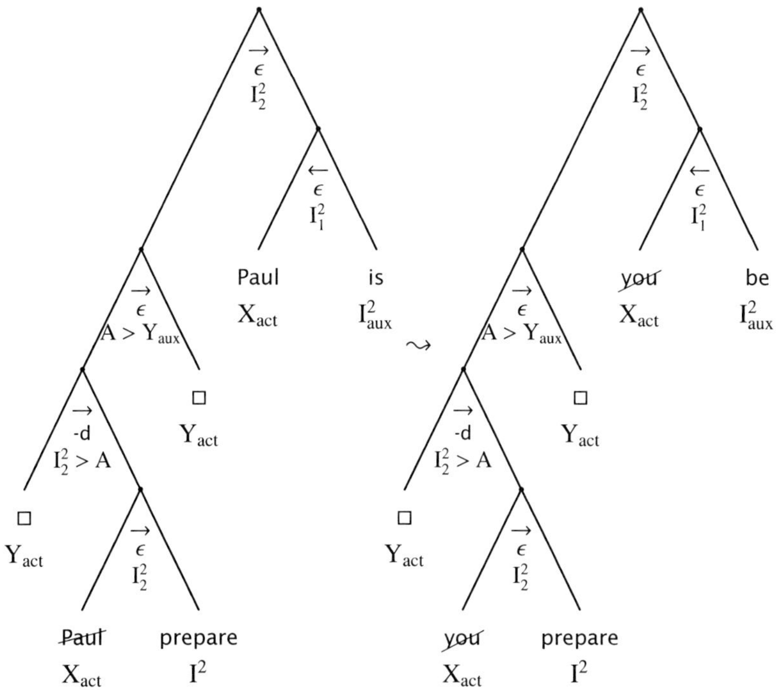

Figure 4.13: The adtrees of *{Paul is / Be} prepared{. / !}*

Finally, with the imperative construction (4-4f) also the first actant become undefined, in semantic terms: in fact, the imperative construction of the 2nd PERSON simply omits the referred person, preferring to follow a pragmatic solution instead. In this particular construction, it has no sense to introduce an extra-valent argument through *by* to reprise the first actant, as it is already in charge.

Why we say this construction, with the advancement of the first actant, is particular? Because the list of verbs that admit this semantic-driven advancement of the first actant is very limited: Alexander (1988) lists only *be prepared!*, *be seated!* and *be warned!*; in fact, in colloquial English the same auxiliary construction can be used, but only with few other verbs, such as *to become* or *to get*: *get dressed! get married! get divorced! get arrested! get killed* and *become concerned!*. The semantic limitations in applying a construction can be represented by semantic traits accordingly.

There are more consequences in the passive than what suspected at a first glance, in particular about the information prominence for the actants of the meaningful construction passivized in the *esse*-construction—see again Section 3.2 about the dichotomy ergativeness *vs.* accusativeness if needed. Let us take the following examples in English (Blake, 1990, with modifications):

(4-5a.) The shah slept in the beds.

(4-6a.) A camel stepped on the package.

The constructions *X slept in Y* and *X stepped on Y* allow passivization as their first actant is in a relation of dependency (→). These are the correspondent complex phrases (the adtrees are in Figures 4.14 and 4.15):

(4-5b.)	Y_{act}	ToBe	$(\overrightarrow{I} > A) + adp$	*by*	$\overrightarrow{X_{act}}$
	The beds	were	slept in	by	the shah.

(4-6b.) The package was stepped on by a camel.

There is another class of bivalent verbs where this transformation is not allowed (this is valid in English but not only), since the trajectory of information prominence of the first actant is reversed. Here are the examples (Blake, 1990):

(4-7a.) The ice cube melted in the oven.

(4-8a.) The toothpaste oozed into the sink.

(4-9a.) Little Red Riding Hood vanished in the woods.

If the transformation chain shown by Figure 4.4 is applied, the result is ungrammatical—signed by the asterisk (*):

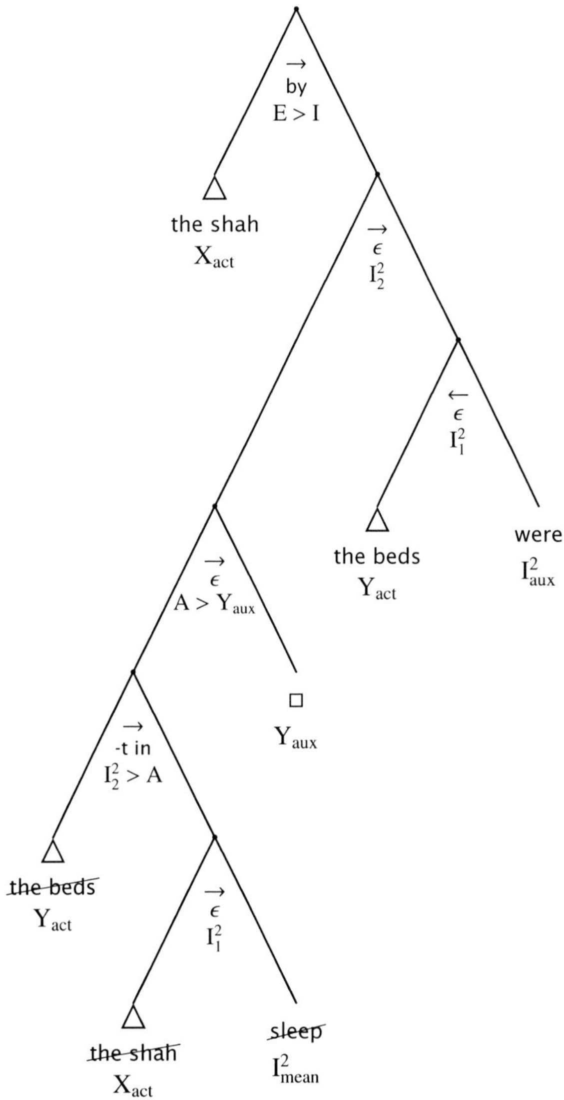

Figure 4.14: The adtree of *...were slept...*

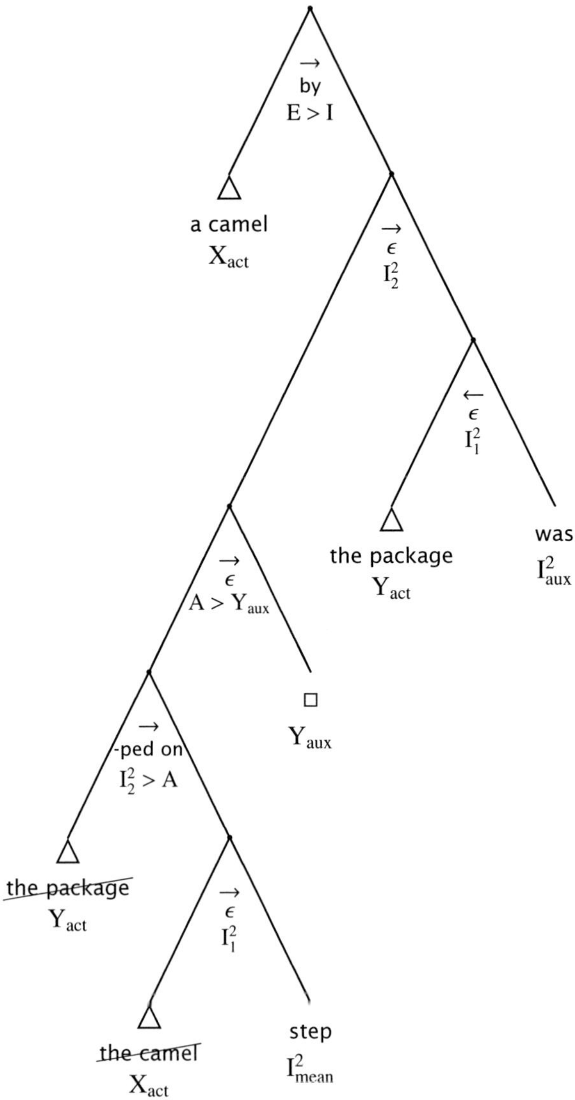

Figure 4.15: The adtree of *...was stepped...*

(4-7b.) $*(Y \quad \text{ToBe} \quad (\overleftarrow{I} > A) + adp \quad by \quad \overleftarrow{X})$
*The oven was melted in by the ice cube.

(4-8b.) *The sink was oozed into by the toothpaste.

(4-9b.) *The woods were vanished in by Little Red Riding Hood.

The active-passive transformation is a good test of the information prominence configuration of the construction in analysis: if it is possible, the stative-verbant relation of the first actant will be a dependency ($\overrightarrow{X}$); otherwise, it will be a government ($\overleftarrow{X}$). The trajectories of the other actants, if any, can be found throughout the explicit application of the active-passive transformation as well.

Information prominence and normalisation

The active-passive transformation has indirect consequences, i.e., not only to the stative-verbant *relation*, but also to the verbant-to-stative *transformation*. In fact, every language applies some strategies in order to use a verbant as a stative. For instance, Horn (1980) already noted that in English the affixes *re-*, *un-*, *-able* and *-ee* can be used with unaccusative verbs (or bivalent verbs having the first stative-verbant relation as a dependency) but not with unergative ones (nor with bivalent verbs having the first stative-verbant relation as a government): [18]

bivalent verbs (I^2) with $\overleftarrow{X}$	*unaccusative* ($\overleftarrow{I}$)	*unergative* ($\overrightarrow{X}$)
re-paint	re-appear	*re-sneeze
re-wash	un-fold	*sneez-able
wash-able	perish-able	*un-swim
draft-ee	absent-ee	*danc-ee

Usually, the transformations—either from verbants, adjuncts, or circumstantials—into statives are referred in the literature as *normalisations*, another common case of grammaticalization. Constructive adpositional grammars show that some information of the predication that is behind normalisation is not lost but only *hidden*: the admissible normalisations derived from certain affixes require unaccusative hidden predications, otherwise they do not work at all. The power of grammaticalization cannot be underestimated: paraphrasing a classic statement by Givón, grammaticalization is the driving force that transforms yesterday's phraseology in today's morphology and finally into tomorrow's syntax.

Another cue for ergativeness or accusativeness in English is given by the *double* normalisation, i.e., obtained in two steps. In the first step, the verb is nominalised by the suffix *-ing* which act as an adposition imposing its grammar character. In the second step, the result is construed into a stative group by the genitive determiner, if the verb was unaccusative, or by *of*-constructions if it was unergative. We can expand the previous examples in the following way:

(4-5c.) Shah's sleeping.

(4-6c.) The camel's stepping.

(4-7c.) The melting of the ice cube.

(4-8c.) The ooze of the toothpaste.

(4-9c.) The vanishing of Little Red Riding Hood.

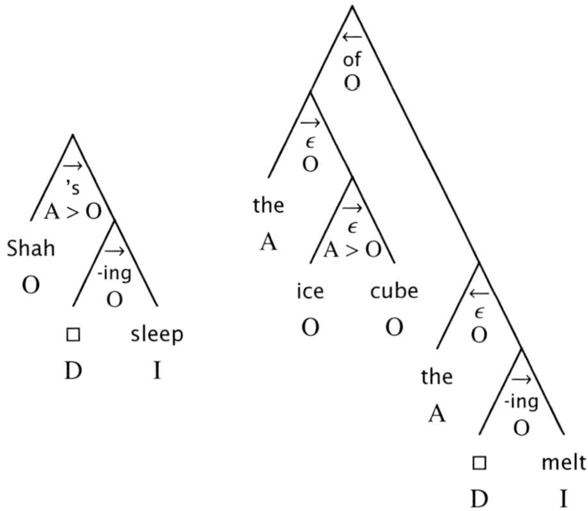

Figure 4.16: Trees of *Shah's sleeping* and *The melting of the ice cube*

It is worth noticing, that this dichotomy 'genitive determiner construction' *vs.* '*of*-construction' is only a tendency, not a general rule of English—it stands on a statistical, non-deterministic basis. What we wanted to show here was the fact that a strong correlation exists between the information prominence standing in the stative-verbant relation and the admitted transformations into statives.

Very often adjunctive-based relations can be nested one into the other recursively: the trees describing the structure of (4-7,9c) are good examples of it. Figure 4.16 shows the trees of (4-7c, left) and (4-9c, right), while the other ones, quite similar, are left to the reader. In (4-7c), the Saxon genitive *Shah's* is a construction that takes the stative grammar character of *Shah* and treats it as if it were adjunctive. Moreover, it request another stative as the governor, in this case *sleeping*—a stative resulted from grammar character imposition.

By contrast, in (4-9c) an example of the so-called 'noun-plus-noun' English construction with *ice cube*, which is very similar to Saxon genitive—the only difference is being the absence of the adposition marker (compare with *Shah's sleeping*). On the right, *melting* is similar to *sleeping*, while the article-stative relation *the melting* is similar to the example *the elephant* (3-23a, Figure 3.20). The indicators in the final adtrees are due to the original stative-verbant constructions that gives raise to normalisation, as we have previously explained.

Presentatives

One of the most important function of natural language is deixis, i.e., the function of referring extra-linguistic entities, or entities out of the joint attentional frame. In order to accomplish this very special function, all languages have developed a set of highly grammaticalized constructions devoted to deixis.

In English—as well as in many languages belonging to the Standard Average European Sprachbund—there are a lot of deictic constructions where the verb *to be* is involved. It is worth noticing that such constructions arise very early during first language learning (Tomasello, 2003). For instance, a standard construction can be used for identification purposes (4-10ab).

(4-10a.) It's a car (not a banana).

(4-10b.) That's the elephant (not the other one, there in the zoo).

Another common purpose of the standard *esse*-construction in children's speech is possession (4-11ab).

(4-11a.) That's my mum (not yours).

(4-11b.) It's Paul's brother (not Charles').

It is worth noticing, that the most prominent information is on the second actant, as we already have shown in (4-3). However, there are special constructions that can configure these trajectories differently. In particular, *there*-constructions are frequent in early child language (Lakoff, 1997). [19]

(4-11c.) There are my shoes (not my socks).

(4-11d.) There is Liza (not Paul).

(4-12a.) My shoes are there (not in the garden).

(4-12b.) Liza is there (not at home).

Analogous constructions can be put at work through the use of *here*:

(4-11e.) Here's a gift (not a cake).

(4-12c.) Paul is here (not in his office).

Let the constructions exemplified in (4-11) be *external -ere*-constructions, while the constructions in (4-12) will be *internal -ere*-constructions, by contrast. The terms 'external' and 'internal' refer to the proximity of the circumstantial group ending in *-ere* to the verbant.

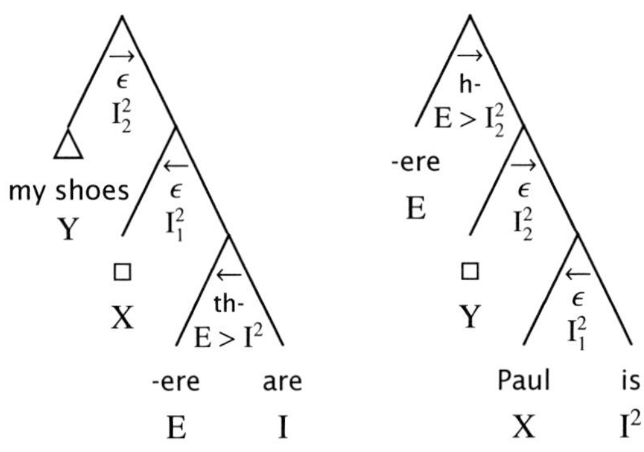

Figure 4.17: Adtrees of *There are my shoes* and *Paul is here*

In Figure 4.17 the reader can see the adpositional trees of internal *-ere*-constructions (4-11c, left) and external ones (4-12c, right). The difference between the two constructions is in the trajectory of information prominence. In fact, in the internal constructions the most prominent group is the second actant (Y), while in the external ones the most prominent is the circumstantial, that takes the collocation of the second actant in the syntagmatic axis.

The morphemic analysis of the circumstantials *th-ere* and *h-ere* will be explained in Section 4.6, devoted to correlatives.

Constructions based on *habere*

The verb *habere* is a luxury that not many languages of the world inherited from Latin. Even within Standard Average European, most Slavic languages do not show auxiliary *habere*-constructions. This is even more true out of that Sprachbund: for instance, Turkish use *var*-constructions and *yok*-constructions to express respectively ownership and lack of ownership.

(4-13.)	kedi	-m	var.
	cat	*my*	*be-there*

Figure 4.18 shows the adpositional tree of the Turkish phrase *kedim var*, usually (and rightly) translated in English as ‘I have got a cat’. But there is no *habere*-construction here, rather a ‘be-there’ (*-var*) with a morpheme marking POSSESSION and PERSON: 1st, SINGULAR (*-m*).

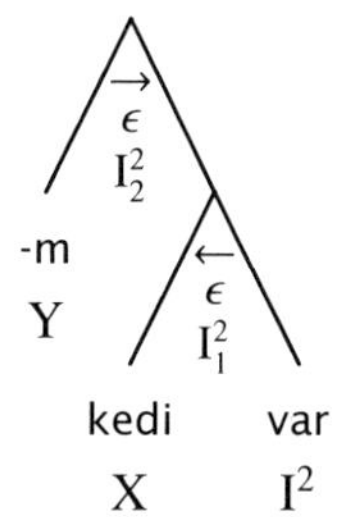

Figure 4.18: Adtree of Turkish *kedim var*

In English grammars, the proper use of *to have*, meaning POSSESS is distinguished by the auxiliary constructions (4-14d), see (Alexander, 1988, 198–199).

(4-14a.) Paul has a Bentley.

(4-14b.) Paul has got a Bentley.

(4-14c.) Paul has got a Ford.

(4-14d.) Paul has washed his car.

In our perspective, the usage of *to have* either properly or auxiliary raises very similar constructions indeed. It is interesting to note that in British English the construction involving the perfect form of *to get* (4-14b) is equivalent to the plain form (4-14a), while in American English (4-14b) is a marked form, where the possession by the American Paul of his Ford is PERFECT, i.e., perhaps Paul now does not possess a Ford anymore.

The reader is invited to notice that the adpositional trees of the *have-got*-construction (4-14b) and the standard auxiliary *habere*-construction (4-14d) are pretty the same—see Figure 4.19, left. It is worth noticing that the adjectivation performed by the PARTICIPLE construction activated by *-ed* the aspect PERFECT (i.e., completed action) and therefore it is a predicative adjective (→). Nonetheless, as the auxiliary second actant (Y_{aux}) is omitted, this fact is not perceived by the speaker—the opposite would be true if the sentence were *Paul has his car washed*: this last case we will see in Section 4.6, devoted to correlatives.

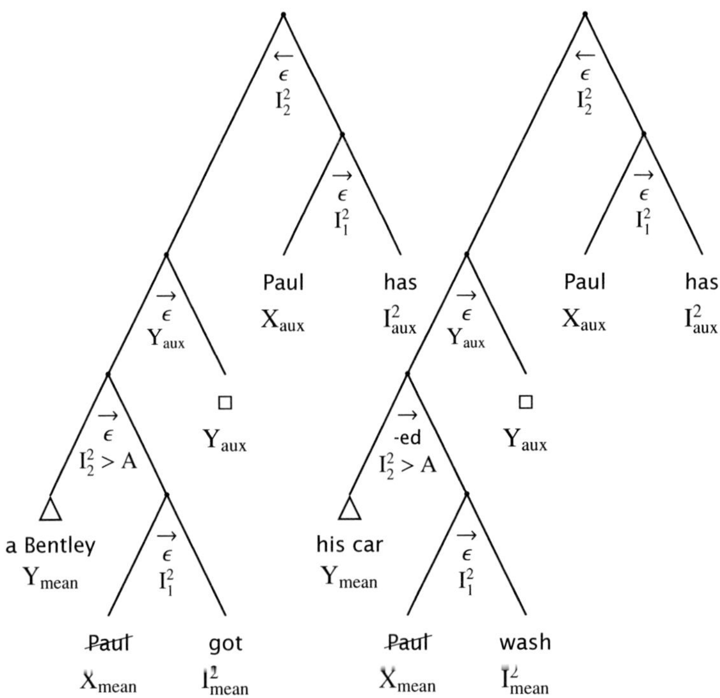

Figure 4.19: The adtrees of *Paul has got a Bentley* and *Paul has washed his car*

As we have already seen about auxiliary *esse*-constructions, also auxiliary *habere*-constructions are complex, as they involve phrases turned into statives, i.e., hidden predications—see *got* (4-14c) and *washed* (4-14d). The second actant belonging to the auxiliary construction (Y_{aux}) is omitted, but the fact that an adjunctive predication is at work is not: that is why the adtree signs it explicitly. These hidden predications have got explicit in (4-14ef).

(4-14e.) $\overrightarrow{X_{act}}$ I^2_{mean} $\overleftarrow{Y_{act}}$
Paul got a Ford.

(4-14f.) $\overrightarrow{X_{act}}$ I^2_{mean} $\overleftarrow{Y_{act}}$
Paul washed his car.

The reader is invited to note that the structure of the predications being hidden is still recognisable. In fact, the *habere*-constructions in English retain the second actant of the active construction of the meaningful verb (Y_{act}), while its first actant is omitted, being the same of the *habere*-construction.

In Romance languages—such as Italian—the choice between *esse*- and *habere*-constructions within the same meaningful verb has some interesting and non immediately evident consequences.

(4-15a.) Paola ha corso la maratona entro quattro ore.

(4-15b.) Paola è corsa a casa.

(4-15c.) *Paola è corsa la maratona entro quattro ore.

(4-15d.) ??Paola ha corsa la maratona entro quattro ore.

In (4-15ab) a constructive pair of examples is shown: the meaningful verbant is a past participle of the verb 'to run' (*correre*). The second actant of the meaningful verb can be expressed only if the proper *habere*-construction is at work, otherwise the result is strictly ungrammatical (4-15c). Moreover, within the *habere*-construction normally the past participle *corso* does not agree with the gender of the first actant—that is why (4-15d) is not acceptable, except perhaps in case of really substandard, marked registers of the Italian language.

The adtrees of (4-15ab) are quite different—compare Figure 4.20 and 4.21. Some observations can be done. Even if *Paola* is a proper name FEMMINILE ('feminine'), it does not agree with the PARTICIPIO ('participle') in the *habere*-construction (4-15a), while agreement is required in the *esse*-construction (4-15b). Furthermore, the structure of the hidden predication has a relevant element of difference: in fact, the second actant of the meaningful verb (Y_{mean}) can be expressed (*la maratona*) or omitted (□) in (4-15a).

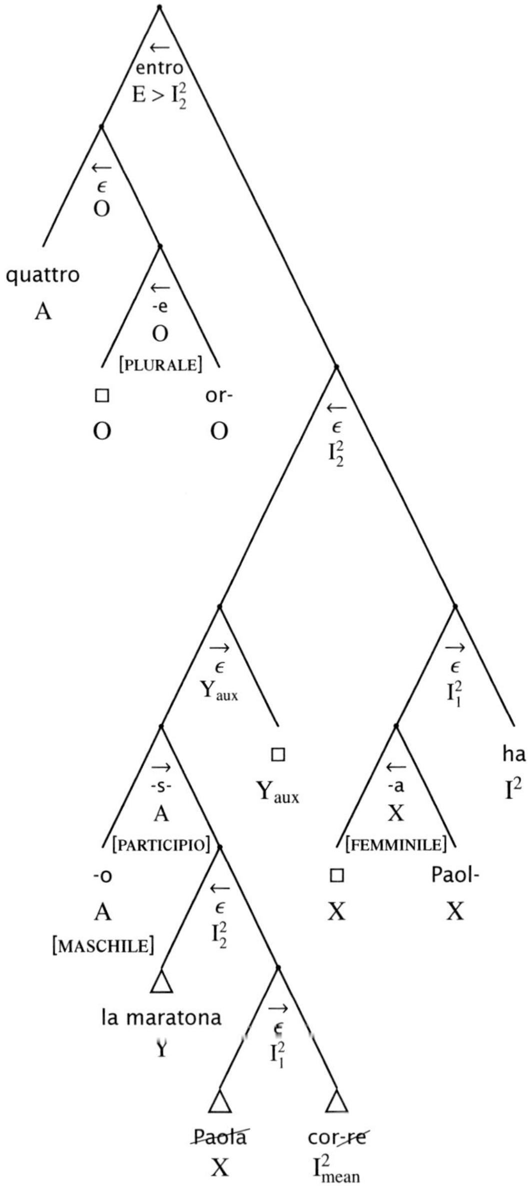

Figure 4.20: The adtree of *Paola ha corso la maratona entro quattro ore*

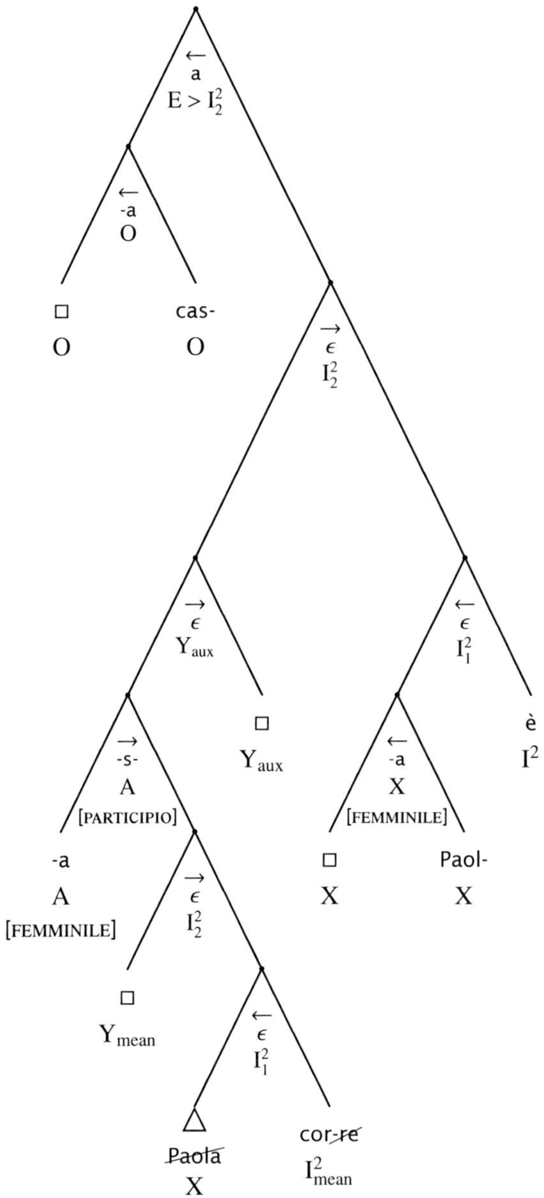

Figure 4.21: The adtree of *Paola è corsa a casa*

As a side note, (4-15a) reveals how the numeral *quattro* ('four') requires the activation of the morpheme *-e* which pluralises the feminine morpheme *or-* ('hour'). In order to help the reader, we have rewritten (4-15ab) with a special emphasis on the main differences:

	$\overrightarrow{X}$		$\overleftarrow{Y}$	
(4-15a.)	Paol-**a**	ha cors-**o**	la maratona	entro quattro ore.
	Pauline	*run*	*within four hours*	

	$\overleftarrow{X}$		
(4-12b.)	Paol-**a**	è cors-**a**	a casa.
	Pauline	*run*	*home*

The dichotomy 'attributive *vs.* predicative' adjectives still rules within Italian *habere*-constructions.

(4-16a.) Mario ha lavato la camicia. (*M. has washed his shirt.*)

(4-16b.) Mario ha la camicia lavata. (*M. has his shirt got washed.*)

Figure 4.22 shows the relevant differences between (4-16a), the standard, unmarked *habere*-construction, and the marked construction, i.e., with the hidden predication put in emphasis with a different collocation: the dependency (→) between the stative *la camicia* and the participle *lavat-{o/a}*, together with agreement, are clear signs of the presence of different constructions at work. In particular, (4-16b) hides a correlative predication—see Section 4.6 below.

4.5 Hidden predications

Natural languages have a lot of constructions in order to hide predications, notably predicative adjectives, infinitives, gerundives and so on. We want to stress another time the fact that when hiding occurs, the hidden information can always be reconstructed if needed.

In the case of hidden predications, this means the adtree of the hidden predication itself is the result of a transformation chain starting from the adtree which describes the corresponding, default, unhidden, explicit phrase.

In this section we briefly present some case studies, useful to understand how to treat these phenomena in general.

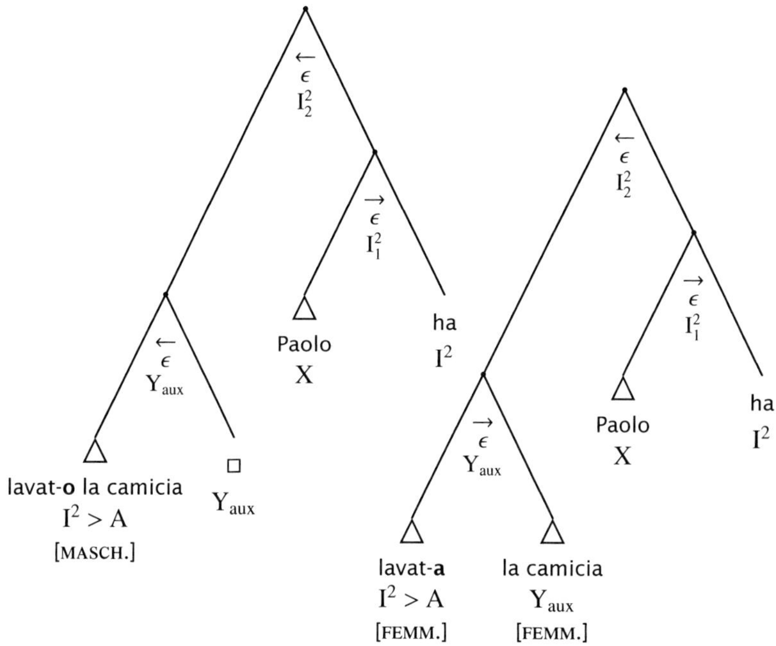

Figure 4.22: The adtrees of *Paolo ha {lavato la camicia / la camicia lavata}*

-ing-constructions and other gerunds

The English *-ing*-constructions are "half-verb and half-noun, which makes them a serious challenge for any theory of grammatical structure" (Hudson, 2007, 184). The challenge is how to treat this sort of 'double nature'. Hudson (2007) uses examples from contemporary and Old English in order to deal with *-ing*-constructions, and he states:

> If we think of a gerund in terms of the phrase that it heads, the following generalisation is almost true: (157) A phrase headed by a gerund is (a) *an ordinary clause as far as its internal structure is concerned, but* (b) *an ordinary noun phrase (or DP) in terms of its external distribution* (Hudson, 2007, 187, emphasis in the original).

In the terms of constructive adpositional grammars, gerunds in general and English *-ing*-constructions in particular are hidden predication. This implies that, from the point of view of the governing phrase, the gerund plays the role of something different from a predication (I): in practice, the group of morphemes pertaining to the gerund, taken as a whole, plays the role of a stative (O), an adjunctive (A) or a circumstantial (E).

In the case of English *-ing*-constructions, the role played is mainly stative (O), as noted by Hudson (2007).

(4-17.) Meeting people is easy.

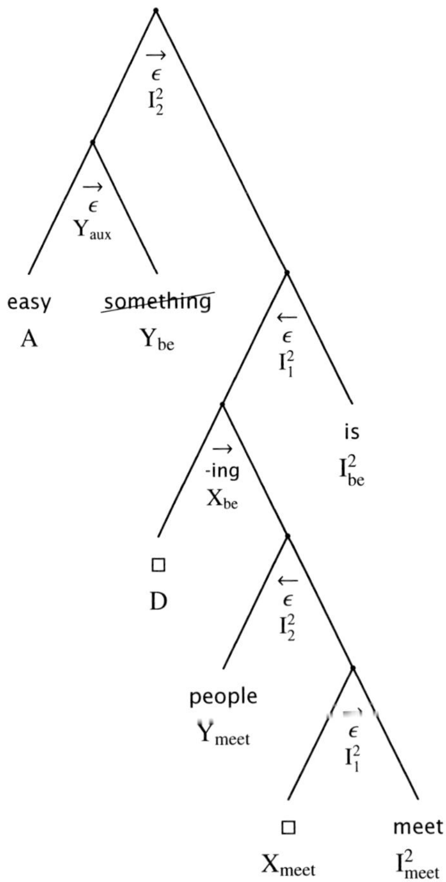

Figure 4.23: The adtree of *Meeting people is easy*

The group of morphemes *meeting people*, taken as a whole, plays the role of the first actant of the verbant *is*, while within its inner structure, *people* is the second actant (i.e. who is met), while the first actant (who meets) is hidden by the *-ing*-construction itself—see Figure 4.23. In order to help the reader, we have used subscripts to distinguish the actants belonging to the *-ing*-constructions, like X_{meet} from the ones belonging to the main, explicit construction, e.g. Y_{be}.

The *-ing*-construction can also play the role of the second actant: in this case the construction is more complex, and in fact the possible activating English verbs are few, in other words they form a small class in this respect.

(4-18a.) The machine needs fixing.

(4-18b.) My house requires painting.

In (4-18) there are two examples of such constructions, with *needs* and *requires* the activating verbant. Their respective adtrees show the same structure.

In Figure 4.24 the adtree of (4-18a) is represented. It is obvious why this construction is used: the goal is to cut off the actant who does really fix the machine; in fact, it existence (X_{fix}) is well hidden in the adtree. Also, there is a need to avoid the repetition of *the machine* as a second actant for *fix*. We will deal with goals such these in Chapter 6, devoted to constructive pragmatics.

The sign $I_2^2 > Y_{need}$ is a compact form to indicate that the predication of *fix* is hidden as a stative thanks to the use of the adposition *-ing*.

English *-ing*-constructions can also play the role of adjunctives, but only as the result of a further grammar character change, from stative to adjunctive, denoted by I > A.

(4-19.) Paul is a working class hero.

Figure 4.25 shows the adpositional tree of the second actant of the *esse*-construction, as the rest does not show anything new—the complete adtree being left to the reader. The gerund *working* is a modifier of *class*, and then *working class* modifies the governor *hero*.

Paraphrasing what said by Hudson (2007), the English *-ing*-construction is a hidden predication playing the role of a stative, but here this resulting stative is again changed in its grammar character by a further construction. In fact, this one is another example of the so-called ‘noun-plus-noun’ construction, a sort of collocational Saxon genitive, where the marker *’s* got lost; we have already seen such construction—see again Figure 4.16 if needed.

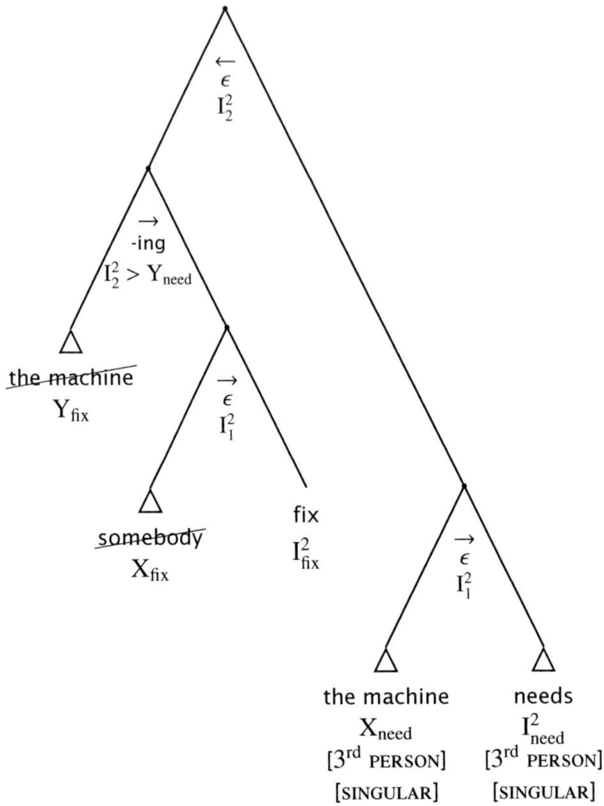

Figure 4.24: The adtree of *The machine needs fixing*

English *-ing*-constructions can play the role of attributive adjunctives, as just seen, or even predicative.

(4-20a.) Paul is a dead man walking.

(4-20b.) ??Paul is a walking dead man.

In (4-20a) we see another particular *-ing*-construction. In fact, the gerund *walking* is put after the governor *man*. This collocation is a sign of adjunctive predication—recall the discussion of the relics of Latin within languages being in contact or directly descendants, like Portuguese: in particular, see again (3-16ab) in Section 3.3, if needed. This fact is tested through (4-20b), which is

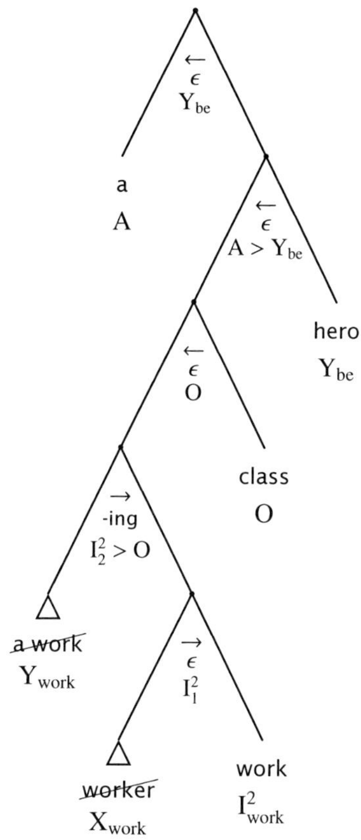

Figure 4.25: The adtree of *...a working class hero.*

hardly acceptable within the same scene, where the same structure—and, hence, adtree—of (4-19) was put at work.

The group *dead man walking* is an idiomatic expression and could be considered as completely unproductive, frozen, i.e., which cannot be further grammaticalized. In this case, the adtree will be reduced to a single leaf. But, if we stipulate that the plural *dead men walking* is acceptable, even if not widely used, we can consider this construction still analysable in parts. Figure 4.26 shows exactly this interpretation: *walking* is an adjunctive which modifies the group *dead man*, taken as a whole—in fact, it has no sense that *walking* modifies *man* alone, if we consider the meaning of the whole expression. The collocation of

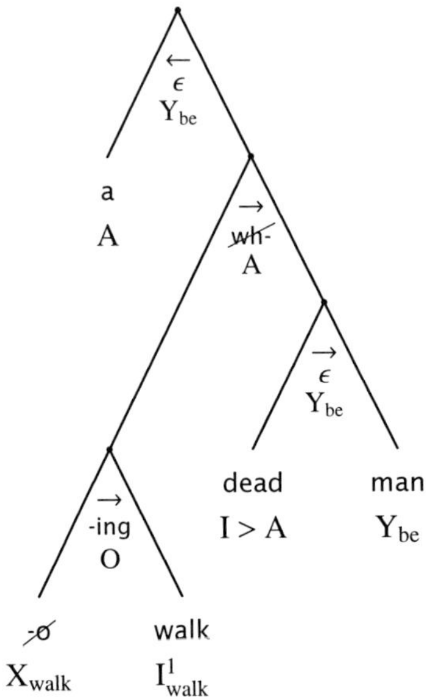

Figure 4.26: The adtree of ...*a dead man walking*.

walking indicates that the gerund, playing the role of a stative in itself, becomes a predicative adjunctive ($\overrightarrow{A}$). What is hidden, here, is a correlative construction: we will see this family of constructions in details in Section 4.6 below.

Finally, an English group headed by the adposition *-ing* can also play the role of a circumstantial (E). In this last possible case, the head of the group will be an appropriate preposition.

(4-21.) Without telling him, Liza cannot be happy.

Figure 4.27 shows the adpositional tree of the circumstantial ...*without telling him*. It is quite obvious that the gerund *telling him* is a stative (compare with the phrase *Liza cannot be happy without Paul*): what is relevant here is the fact that the hidden verbant *tell* is trivalent, and that is why the adposition *-ing* heads such a rich adtree. As a final observation, the reader is invited to note the fact that the English preposition *without* is the result of the grammaticalization of the Old English *withŭtan*, i.e., 'in the out side of'.

Not all gerunds in all languages share the features seen for English. For instance, in Italian, gerunds play the role of circumstantials directly, not of sta-

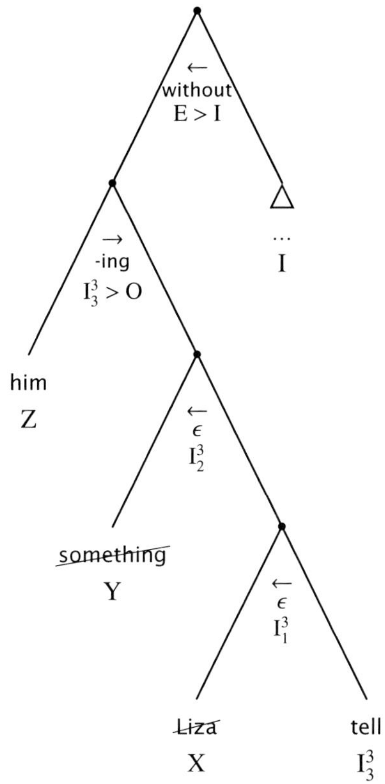

Figure 4.27: The adtree of *...without telling him.*

tives—for this last role, Italian use infinitive forms. Furthermore, the trajectory of information prominence of the hidden verbant leads the language to choose between an *esse*- or *habere*-construction, choice that has a lot of consequences.

(4-22a.) Avendo telefonato Lisa, Paolo pianse.
(*After the phone call by Liza, Paul cried.*)

(4-22b.) *telefonato Lisa, Paolo pianse.

(4-22c.) Essendo partita Lisa, Paolo pianse.
(*After the leaving by Liza, Paul cried.*)

(4-22d.) Partita Lisa, Paolo pianse.
(*After the leaving by Liza, Paul cried.*)

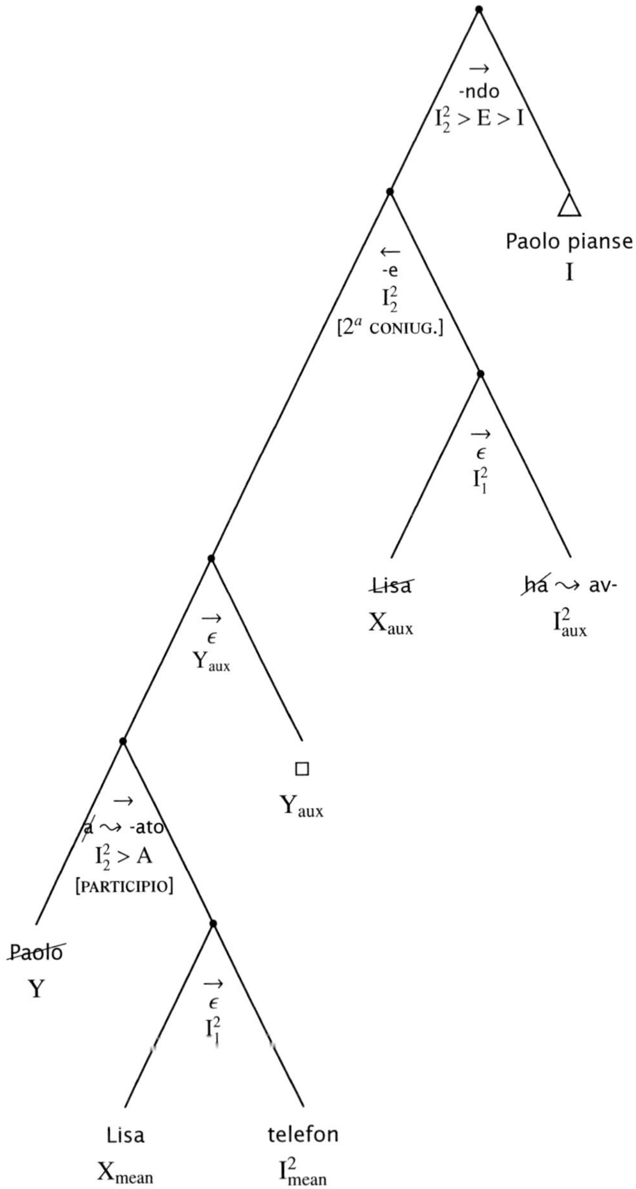

Figure 4.28: The adtree of *Avendo telefonato...*

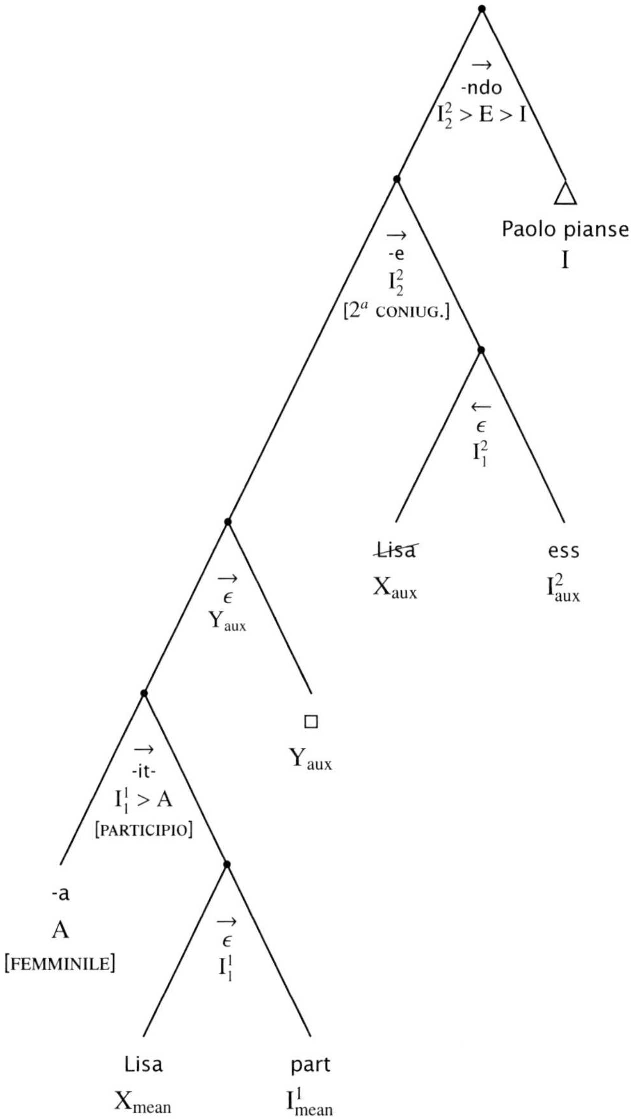

Figure 4.29: The adtree of *Essendo partita...*

Infinitive, a luxury

The infinitive construction is a luxury few languages in the world permit themselves: for instance, within the Balkan Sprachbund (roughly, from Albanian to Romanian), it is lost, even if these languages belongs to the Standard Average European (Comrie, 1989, 207). In Romance languages, the infinitive of Latin survives in different forms. In most cases, the role played by the infinitive is to use the verb as a stative, while some or all actants are hidden—the same role of English *-ing*-construction just seen, as gerunds play the role of circumstantials.

(4-23a.)	A Lisa	piace	ballare.
	to-Liza	*like*-3rd PERS: PLUR	*dance*-INF.

(4-23b.)	*A Lisa	piacciono	ballare.
	to-Liza	*like*-3rd PERS: PLUR	*dance*-INF.

(4-23c.)	*A Lisa	piaci	ballare.
	to-Liza	*like*-2nd PERS: SING	*dance*-INF.

(4-24.)	Paolo	vuole	provare	ad andare	a fare	una vacanza.
	Paul	*wants*	*try*-INF	*go*-INF	*do*-INF	*a holiday.*

In Italian, *piace* (4-23) activates a construction reversed compared to the English *like*: the first actant is the infinitive (*ballare*), while what is a subject in English is an oblique—a sort of dative—in Italian, because of the presence of the preposition *a*. This example shows that the infinitive can play the role of whatever actant, even the first one (Figure 4.26). (4-23bc) shows that the Italian infinitive is 3ª PERSONA: SINGOLARE ('3rd person: singular'). As we will see below, other Romance languages mark the person of the infinitive explicitly.

In Italian it is possible to nest an infinitive into another infinitive recursively. (4-24) shows a rather extreme example, but found in colloquial contemporary Italian (Figure 4.31). It is worth noticing that the nested infinitive should be head by the preposition *a* (or *ad*, for phonotactic reasons).

(4-25.)	Lamento	eles	ter	-em	perdido	o comboio.
	I regret	*they*	*have*	INF: 3rd-PERS: PL	*lost*	*the train.*

Sometimes information hiding is only partial. In the so called 'inflected infinitive', the PERSON of the first actant is retained within the infinitive itself: this occurs in Romance languages such as Portuguese, Galician and Sardinian, which are conservative in this respect (Scida, 2004). (4-25) shows an example in Portuguese from Madeira (1994)—see Figure 4.32.

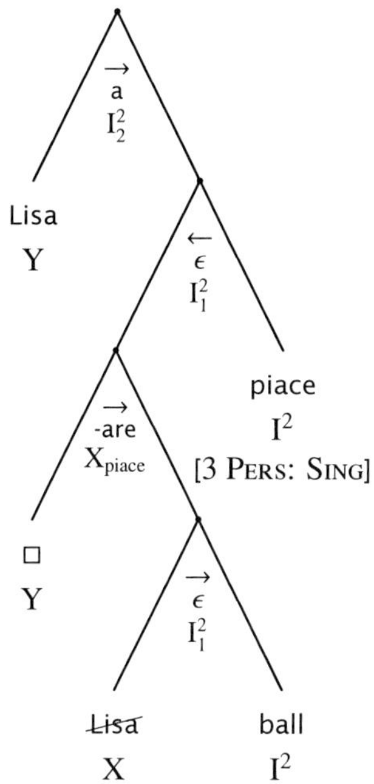

Figure 4.30: The adtree of *A Lisa piace ballare*

What can be done: *-able*-constructions

The *-able*-construction in English, for example *know-ledge-able*—along with its allomorph *-ible*, e.g. in *corrupt-ible*, for phonotactic reasons—was taken from Latin because of the long contact with this language and French (Stein, 2007, 1). It is a good example of hidden predication as an adjective, of course of the predicative type ($I > \overrightarrow{A}$). Its function is to indicate that the second actant in the active correspondent construction (Y_{act}) can be done.

(4-26a.)

Mario	ha	una camicia	lav-abile	a secco.
Mario	*has*	*a shirt*	*wash-able*	*to dry.*

(4-26b.) *Mario ha lavabile una camicia a secco.

(4-26c.) *Mario ha una lavabile camicia a secco.

The Italian sentence (4-26a) means 'Mario has got a shirt which can be dry-cleaned,' *lavare a secco* being a idiomatic construction where the flexible part is

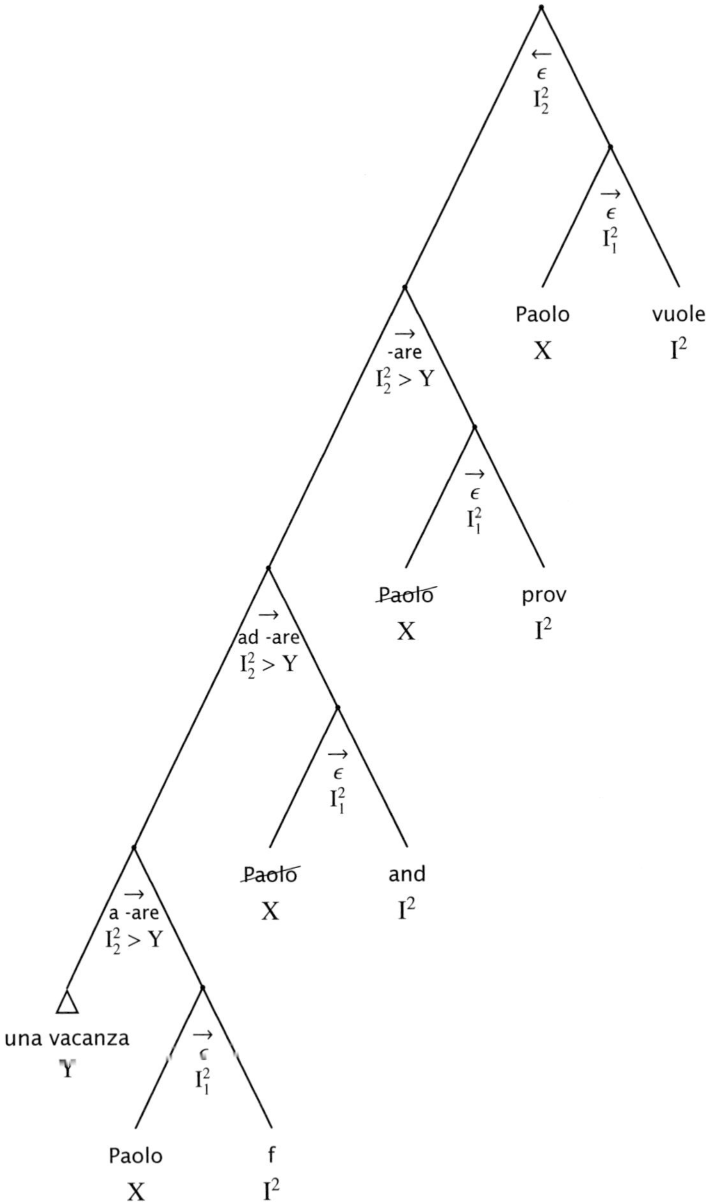

Figure 4.31: The adtree of *P. vuole provare ad andare a fare una vacanza*

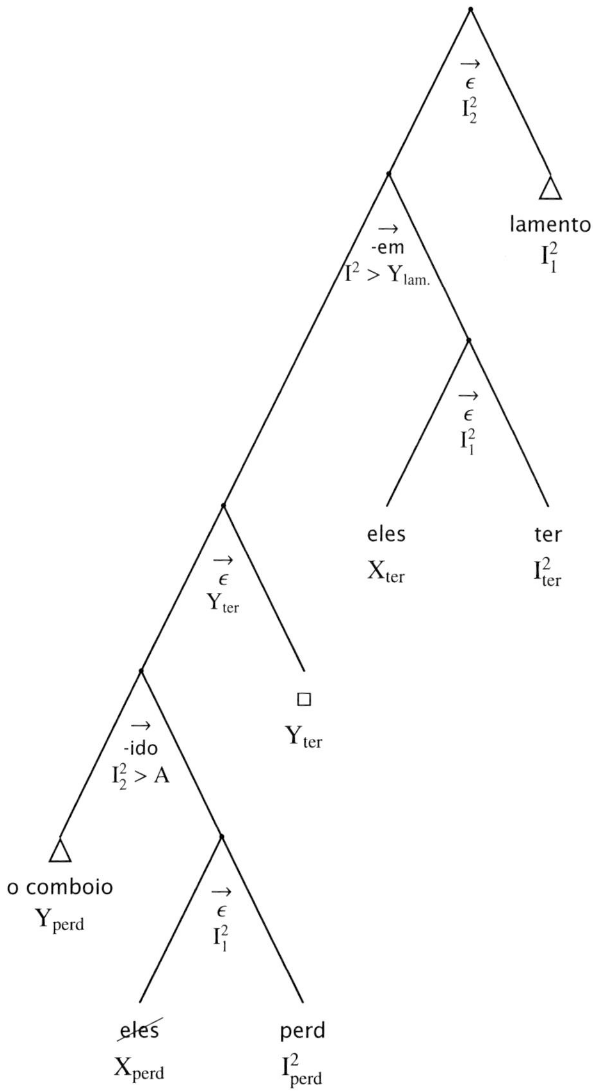

Figure 4.32: The adtree of *Lamento eles terem perdido o comboio*

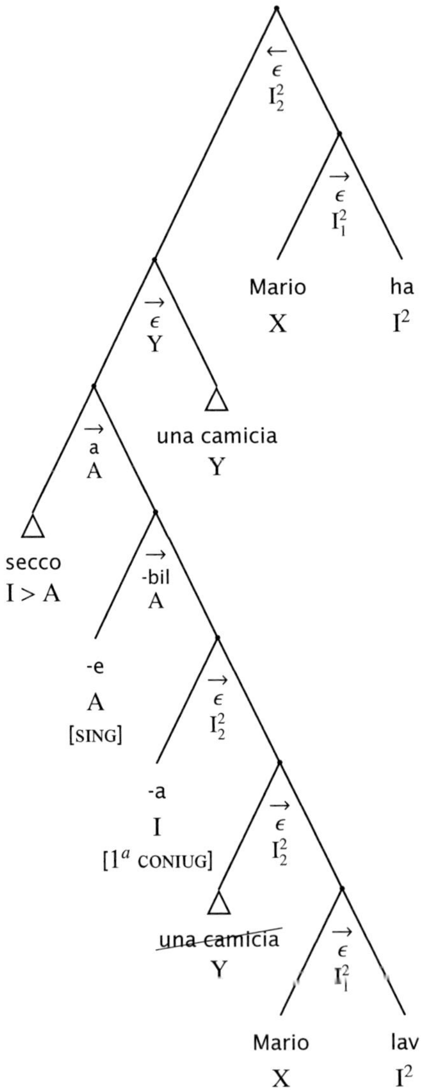

Figure 4.33: The adtree of *Mario ha una camicia lavabile a secco*

the verbant *lavare* ('to wash'), while *a secco* is fixed—*secco* being itself derived from the verb *seccare*, and therefore in the adtree its grammar character is signed appropriately through I > A. (4-26bc) show that the collocation of the Italian construction head by *-abile* is fixed: the adjective comes mandatorily after the noun on which the hidden predication is performed. The correspondent adtree (Figure 4.33) hides a correlative construction—see Section 4.6 below.

The adposition *-abile* is further analysable in Italian, because that language marks the first conjugation and the other ones differently—with *-a* and *-i* respectively, while in English they are relics, completely fixed in the morphemes (*able* and *ible*). Furthermore, the final suffix *-e* is a marker of the singular (SING).

The double nature of the French *de*

The French preposition *de* is a challenge for every student, even if advanced, because of its versatility. Usually French grammars list a plethora of uses on a semantic basis, with some examples, without explaining how to win this challenge of orienteering. The adpositional paradigm can be useful in such cases, because constructions become elicited and some order can be found.

The French preposition *de* has a double nature, according to the construction in which it is put. Essentially, *de* heads a relation where the dependant (*dep*) is always stative, while the governor (*gov*) can be stative or verbant.

(4-27.)	Le livre	de	Pierre.
	the book	*of*	*Peter.*

(4-28.)	Une tasse	de	thé.
	a cup	*of*	*tea.*

(4-29.)	Une salle	de	classe.
	a room	*of*	*class.*

The examples above show that the governor of the group head by *de* is always a stative: *le livre* ('the book', (4-27), which in English corresponds to the Saxon genitive *Peter's book*); *une tasse* ('a cup', (4-28), same construction as in English); *une salle* ('a room', (4-29), which in English corresponds to the collocational Saxon genitive 'noun-plus-noun' *classroom*). The correspondent adtree in Figure 4.34 are quite obvious now: the usual compact notation for attributive adjectivation A > O was substituted with the explicit notation, as the attention here is exactly put on this relation and also as these adtrees are small enough.

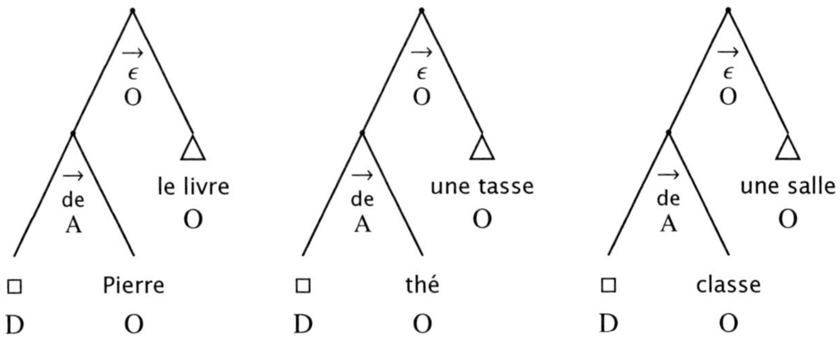

Figure 4.34: Adtrees of French *de* governed by statives

The situation is different if the governor is a verbant: in this case the preposition *de* introduces circumstantials, belonging to various semantic roles, e.g. INSTRUMENT (4-30), MANNER (4-31), PLACE (4-32).

(4-30.)	Je écrire *I write*	de *with*	la main gauche. *the-left-hand.*

(4-31.)	Je répète *I recite*	de *from*	mémoire. *memory.*

(4-32.)	Le train *the train*	(parti) de *(come) from*	Paris… *Paris…*

Please note that the English translations of the preposition *de* in the examples above are indicative: it is not possible to translate the preposition without the concerned stative, because of grammaticalization. For instance, the French *de mémoire* in (4-30) is an idiom like the English constructions *from memory* or *by heart*—please note the different prepositions! However, the English-speaking student of French can be helped here as he becomes aware that this class of *de*-constructions is circumstantial instead of stative, and he knows this fact simply looking at the governor (Figure 4.35, left and middle).

A discourse apart should be done for (4-32), depicted in Figure 4.35, right: here, a *de*-construction is nested into a predicative adjectivation governed by a PARTICIPE: PASSÉ ('participle: past'), a type of construction already seen before. Even if this past participle is not expressed (□), the construction conserves its trace indeed.

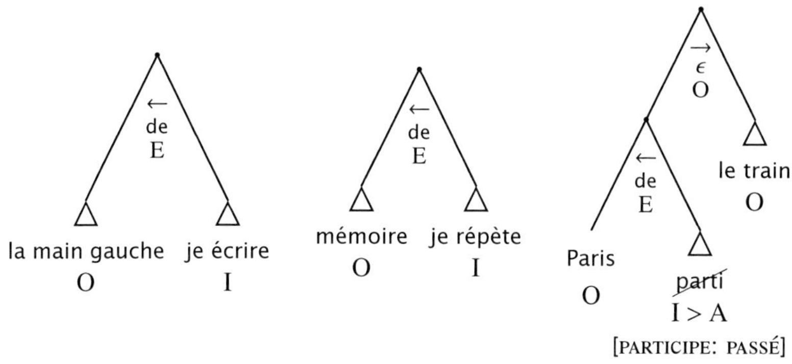

Figure 4.35: Adtrees of French *de* governed by verbants

Resultatives

Resultative are constructions where there is a main phrase, responsible of the actual action, and a hidden predication, where the verbant indicates the result of the actual action. We will sign through appropriate subscripts both the main phrase (e.g., I_{main}) and its result (e.g., I_{result}).

(4-33.) Liza wiped the table clean.

In (4-33) we have an example in English. The main phrase is explicit (*Liza wiped the table*) while the result (*clean*) is the residual morpheme that indicates the result of the hidden phrase *Liza clean the table*.

The adpositional tree of (4-33) is in Figure 4.36. The surviving verbant morpheme *clean* indicating the result becomes a predicative adjective of the second actant of the main phrase *the table*. The adtree shows—under the cancelled morphemes—the structure of the correspondent correlative phrase which is written explicitly in (4-33corr).

(4-33corr.) Liza wiped the table (wh-ich she) clean(-ed).

Resultatives are very powerful constructions: everything inside them gets hidden save what is really needed, i.e., the verbant morpheme turned into a predicative adjective.

(4-34a.)	X	I_{main}	Y	$I_{result} > \overrightarrow{A}$
	Paul	has got	his car	washed.

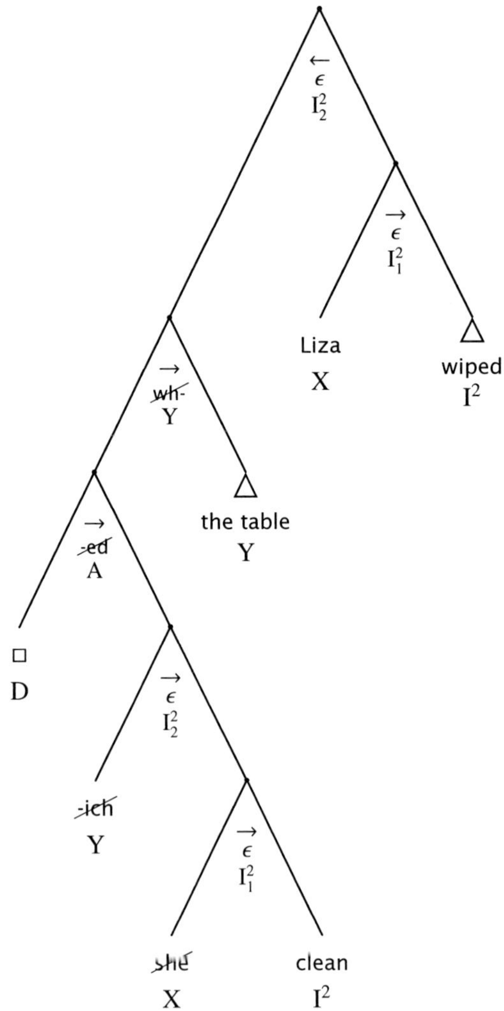

Figure 4.36: The adtree of *Liza wiped the table clean*

(4-34b.)	X	I_{main}	Y	$I_{result} > \overrightarrow{A}$
	He	drove	her	mad.

It is worth noticing that even the so-called ‘pseudo-reflexive’ constructions in reality are only a special case of resultatives, where the hidden predication is governed by an *esse*-construction which got lost, see examples (4-34b). [20]

4.6 The which hunt

Correlatives are all those words that transform a secondary phrase—or even a sentence, i.e., a group of phrases—into a stative, an adjunctive or a circumstantial of the main phrase. In English, some of these words are: *who* (into stative); *which* (into adjunctive); *when* (into circumstantial).

By extension, we will refer to the transformed secondary phrases as ‘correlatives’ too. The term ‘relative’—used in many grammar traditions—in our context indicates a special type of correlatives.

Why languages have developed correlatives? In every language, there are classes of morphemes devoted to put into relation two phrases taken as single leaves. Let the main phrase be the governor (I_{gov}), while the secondary phrase will be the dependent (I_{dep}). They can be classified along their trajectory of information prominence: some leave the prominence underspecified (↔); some others give more prominence to the dependent phrase (→), while, on the contrary, the are some conjunctions that give more prominence to the governor (←). Some examples of English conjunctions per type are shown in Table 4.1.

(4-35a.)	I_{gov}	$\overleftarrow{I}$	I_{dep}
	Alfred can pay	because	he is rich.
	Alfred peut payer	parce que	il est riche.
	Alfred può pagare	perché	è ricco.

(4-35b.)	I_{gov}	$\overrightarrow{I}$	I_{dep}
	Alfred can pay	hence	he is rich.
	Alfred peut payer	donc	il est riche.
	Alfred può pagare	dunque	è ricco.

In (4-35ab) we have provided a minimal pair showing the different information prominence head by a single conjunction (or, in case of French *parce que*,

Table 4.1: Some English conjunctions and their trajectory

trajectory	$\overrightarrow{I}$	$\overleftrightarrow{I}$	$\overleftarrow{I}$
examples	*but*, *if*, *so*, …	*(both) and*, *(either) or*, …	*while*, *however*, …
prominence	I_{dep}	$I_?$	I_{gov}

a grammaticalized expression). It is worth noticing that the structure is the same in English, French and Italian—but this list could be easily extended.

Figures 4.37 and 4.38 show the respective adpositional trees: the only difference is in the inter-phrasal adposition, i.e., in the trajectory signed by the conjunction which connects the two phrases.

Now, it is possible to perform a transformation chain in order to obtain the phrases in (4-36a) and (4-36b), where the less prominent phrase has become an adjunctive of the subject *Alfred* through the use of the correlative *who*, without losing the most important part of the original sentences (4-35a) and (4-35b):

(4-35a.) Alfred can pay because he is rich.

(4-36a.) Alfred, who is rich, can pay.

(4-35b.) Alfred can pay hence he is rich.

(4-36b.) Alfred, who can pay, is rich.

A typological comparison of correlatives

If we look at correlative adposition from a broad typological perspective, we will note that these words are formed by two parts: an adposition—usually the prefix—which is appended to the main phrase (I_{gov}), and does not change, and a flexible part—usually the final—which acts as a pronoun in the correlative phrase (I_{dep}). Following Cicero's taxonomy proposed in *De inventione*, a classic work about rhetorics which highly influenced Western culture (Perelman and Olbrechts-Tyteca, 1958), we can show a typological comparison of the main correlatives (Table 4.2: fixed parts in bold).[21]

Nevertheless, this double nature of the correlative adpositions also stands out of the Western culture. In fact, in Table 4.2 we have introduced both languages

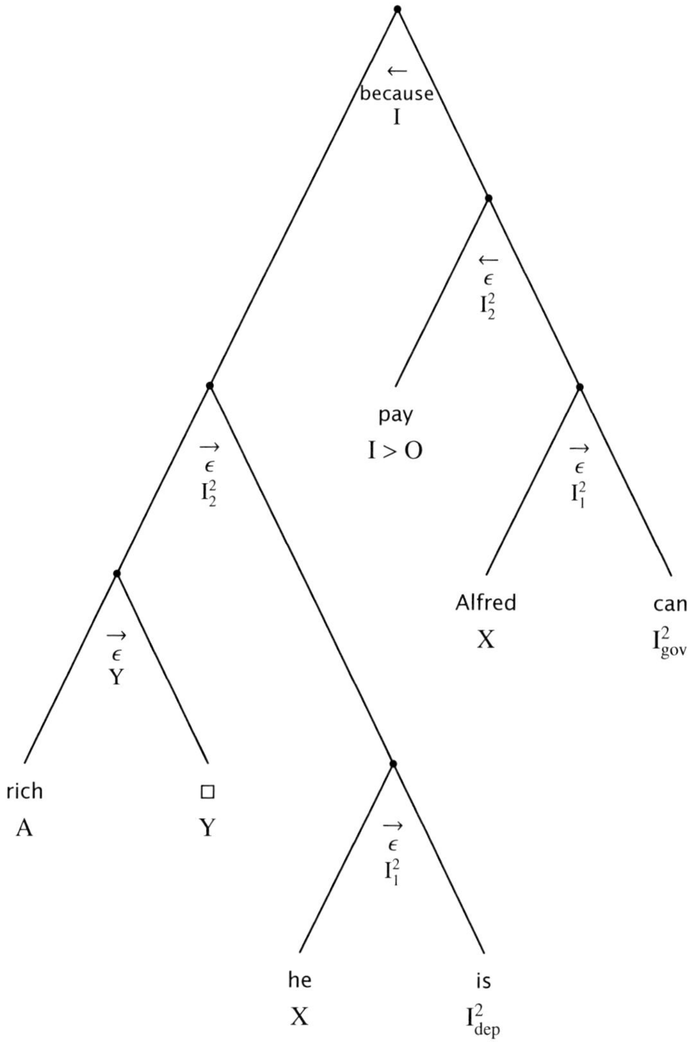

Figure 4.37: The adtree of *Alfred can pay because...*

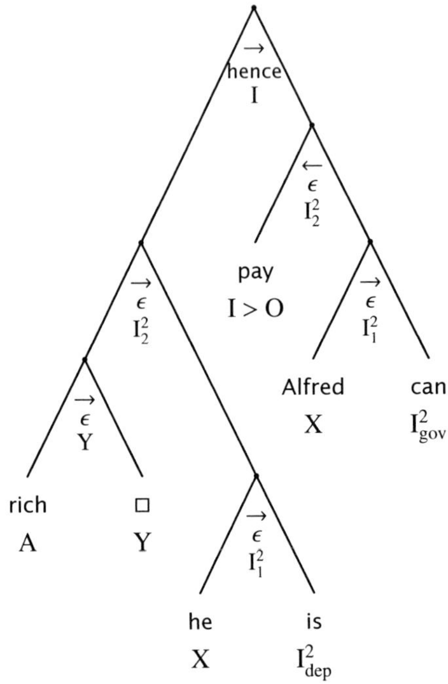

Figure 4.38: The adtree of *Alfred can pay hence...*

belonging to the Standard Average European Sprachbund and languages belonging to other Sprachbund, e.g. Kiswahili belonging to the Bantu, and Turkish belonging to the Turkish.

It is clear that, regardless of the Sprachbund, we can always find a flexible and a fixed part in almost every correlative. We say 'almost' because sometimes grammaticalization—influenced by language policy and planning issues, in particular, the choice of the writing system—can have as a side result to make this structural difference opaque. In those cases, we will put an epsilon accordingly.

A further clarification is needed for the taxonomic element *persona*. In some languages, there is a sememe ANIMATE in action (e.g., in English, *who* for animate and *which* for inanimate). This is the case of English, Italian, Catalan, French, Indonesian, Turkish. In other cases, this distinction simply does not hold, as in German, Dutch, Latin, Polish, Lithuanian and Japanese.

Finally, in Kiswahili there is a subtle distinction in this sense, as the flex-

Table 4.2: A typological comparison of the main correlatives

<table>
<tr><th>Language</th><th>persona</th><th>factum</th><th>causa</th><th>locus</th><th>tempus</th><th>modus</th></tr>
<tr><td>English</td><td>wh-{ich|o}</td><td>wh-at</td><td>wh-y</td><td>wh-ere</td><td>wh-en</td><td>how</td></tr>
<tr><td>German</td><td>w-er</td><td>w-as</td><td>w-arum</td><td>w-o</td><td>w-ann</td><td>w-ie</td></tr>
<tr><td>Dutch</td><td>w-ie</td><td>w-at</td><td>w-aarom</td><td>w-aar</td><td>w-anneer</td><td>hoe</td></tr>
<tr><td>Latin</td><td>qu-is</td><td>qu-id</td><td>{qu-are
|cur}</td><td>ubi</td><td>qu-ando</td><td>qu-o
-modo</td></tr>
<tr><td>Italian</td><td>{qu-ale|chi}</td><td>cosa</td><td>perché</td><td>dove</td><td>qu-ando</td><td>come</td></tr>
<tr><td>Catalan</td><td>{qu-al|qu-i}</td><td>qu-è</td><td>per qu-è</td><td>on</td><td>qu-an</td><td>com</td></tr>
<tr><td>French</td><td>{qu-el|qu-i}</td><td>qu-oi</td><td>pour
qu-oi</td><td>où</td><td>qu-and</td><td>com-
ment</td></tr>
<tr><td>Polish</td><td>k-uris</td><td>k-as</td><td>k-ódel</td><td>kur</td><td>k-ada</td><td>k-aip</td></tr>
<tr><td>Lithuanian</td><td>kto</td><td>co</td><td>dlaczego</td><td>gdzie</td><td>k-iedy</td><td>jak</td></tr>
<tr><td>Indonesian</td><td>{yang mana
|si-apa}</td><td>apa</td><td>meng
-apa</td><td>di
-mana</td><td>k-apa-n</td><td>bet
-apa</td></tr>
<tr><td>Turkish</td><td>{hangi|kim}</td><td>ne</td><td>{niye
|neden
|niçin}</td><td>{nere
|hani}</td><td>ne
zaman</td><td>nasıl</td></tr>
<tr><td>Japanese</td><td>do-no</td><td>do-re</td><td>do
-shite</td><td>do-ko</td><td>itsu</td><td>do
-chira</td></tr>
<tr><td>Kiswahili</td><td>amba
-{ye|yo
|o|lo|cho
|vyo|zo}</td><td>ni-ni</td><td>kwa
ni-ni</td><td>amba
-{po
|ko
|mo}</td><td>wakati</td><td>jinsi
ga-ni</td></tr>
</table>

ible part appended to the fixed *amba* involves not only animateness but also agreement with gender and number, and furthermore it introduces the locative (Toscano and Sewangi, 2005).

In the next sections we will deal with correlatives in details. As already said, the function of a correlative adposition is to transform a dependent phrase (I_{dep}) into an adjunctive (A), circumstantial (E) or stative (O) element.

It is worth noticing that correlative adpositions are predicative, and therefore they retain the information prominence, i.e. their trajectory is of dependency.

Adjunctive correlatives

In Comrie's analysis of relative clause formation, a classic in the literature, we find a set of examples that help us to see how the adpositional trees of adjunctive correlatives are structured. These examples are in English as "English presents essentially no evidence of any kind of restriction, since it is possible to relativize on, for instance, subject, direct object, non-direct object, and possessor in the possessive construction" (Comrie, 1989, 155).

(4-37a.) the man [who bought the book for the girl].

(4-37b.) the book [which the man bought for the girl].

(4-37c.) the girl [for whom the man bought the book].

(4-37d.) the boy [whose book the man bought for the girl].

The verb *to buy* is quadrivalent: in the default active saturated construction the first actant is the Buyer ((4-37a), Figure 4.39), the second one the Object to Be Bought ((4-37b), Figure 4.40), the third one the Seller, the fourth actant being the Price. With a different arrangement, the same actants are valid in verbs like *to sell.*[22] In (4-37c) the third and fourth actants are omitted, while an external circumstantial is inserted, i.e., *the girl*, which express the Benefactive sememe—here, there is a *donato* like predication completely hidden. It is interesting that in English the Benefactive can be referred by the flexible part of the correlative adposition ((4-37c), Figure 4.41), while the Price cannot:

(4-37e.) *the book [for 10 pounds the man bought for the girl].

Furthermore, in (4-37d) the sememe referred by the correlative is the Possessor, which is one of the most important sememes in all languages of the world—see Figure 4.42. (4-37abcd) show that the grammar character of *wh-* belongs to the class of predicative adjectives.

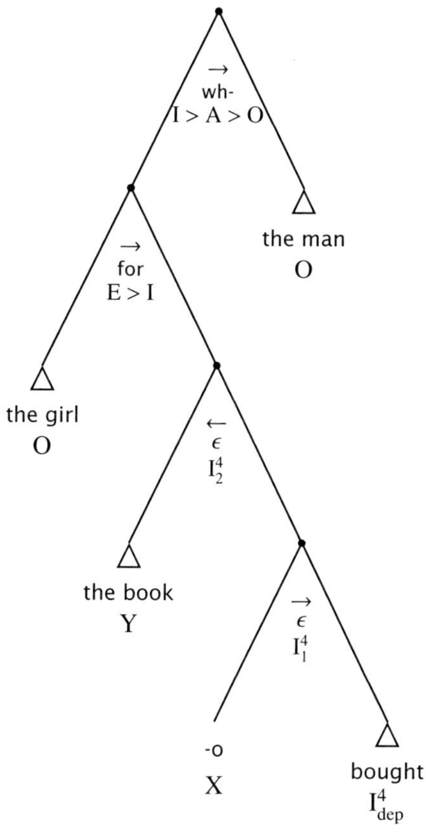

Figure 4.39: The adtree of *the man who bought the book for the girl*

Circumstantial correlatives

Correlatives expressing TIME are more likely to be appended to the main phrase (I_{gov}) as circumstantials, because normally sememes expressing TIME are not part of the in-valent arguments.

(4-38a.) [When Paul will come back] Liza will smile.

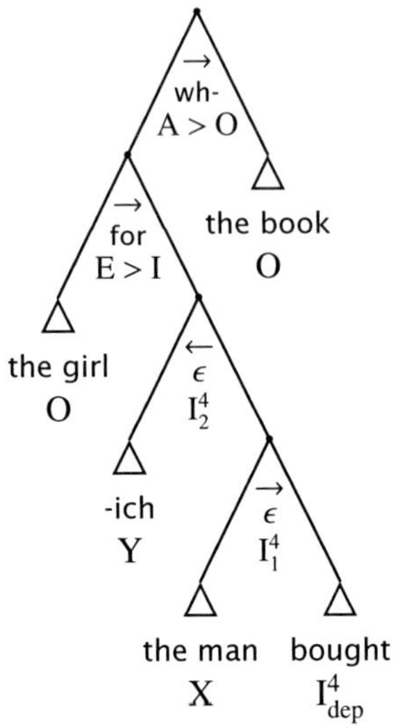

Figure 4.40: The adtree of *the book which the man bought for the girl*

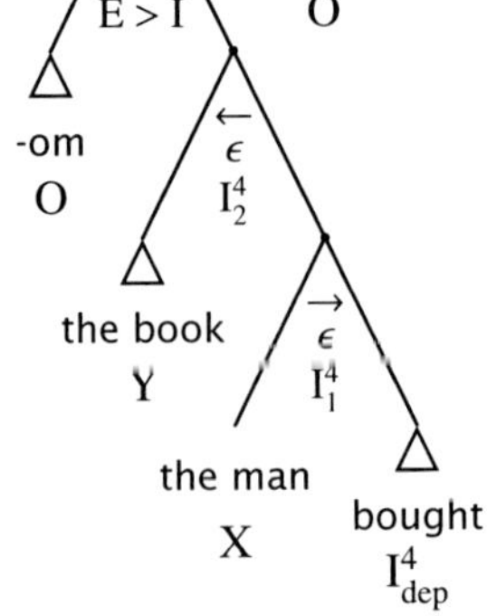

Figure 4.41: The adtree of *the girl for whom the man bought the book*

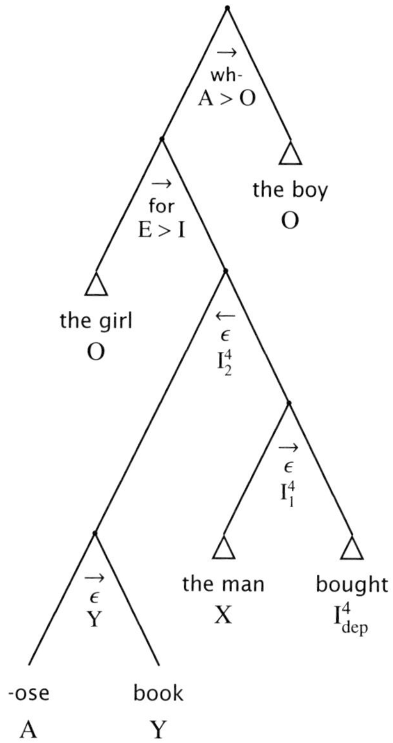

Figure 4.42: The adtree of *the boy whose book the man bought for the girl*

(4-38b.) [Tomorrow] Liza will smile.

Figure 4.43 shows the adpositional tree of (4-38a). From the point of view of the main phrase, the whole correlative *When Paul will come back* is a circumstantial equivalent to *tomorrow* of (4-38b), save for semantics.

Stative correlatives

Sometimes a correlative phrase directly plays the role of an actant. In those cases, the external grammar character of the correlative will be stative, while the internal grammar character can be stative, adjunctive, or circumstantial alike.

(4-39a.) I wonder [how many errors I did].

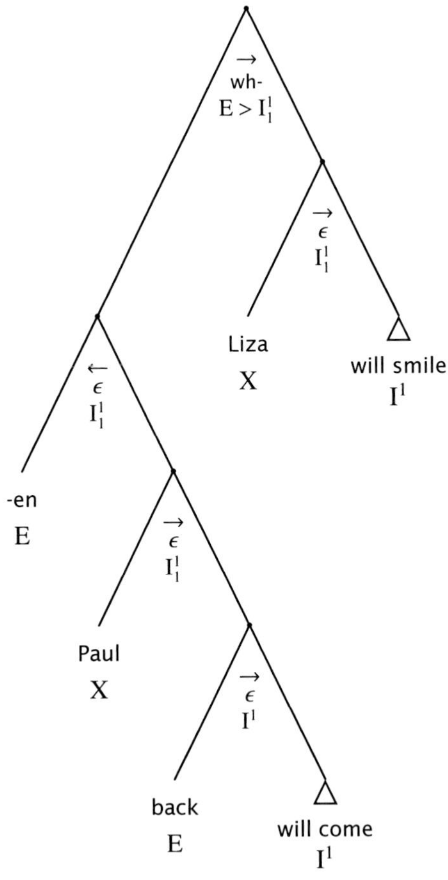

Figure 4.43: The adtree of *When Paul will come back Liza will smile*

(4-39b.) [What time] is it?

For instance, in (4-39a) the internal correlative is a circumstantial-adjunctive relation—see Figure 4.44 and Section 3.5 again, if needed. The trajectories are adjusted accordingly, as the most prominent information is put in *how many*.

In (4-39b) the most prominent actant—the second one, in an *esse*-construction—is taken by the correlative, while the interrogative construction reverse the collocation of the actants themselves (Figure 4.45).

Correlative answers

The typological comparison of correlative adpositions shown in Table 4.2 refers to the activating morphemic group of correlative phrases, which can be considered equivalent to questions.

(4-37f.) Who bought the book for the girl? The man.

(4-38c.) When will Liza smile? When Paul will come back.

(4-39c.) How many errors I did? I wonder.

In many languages there are correlative adpositions devoted to express a deictic answer that relies on the knowledge that the speaker assumes on the receiver. Within the Standard Average European Sprachbund, they are often referred as 'demonstratives' (Alexander, 1988, for English). What is interesting for us is their morphology: the flexible part corresponds to the types used before (i.e., *persona*, *factum*, etc.) while there is a fixed part whose function is exactly to be a correlative answer.

For this reason, in Figure 4.17 (left), which illustrates the adpositional tree of (4-11c), the word *there* was analysed as *th-ere*, *th-* being the fixed correlative answer while *-ere* expresses the LOCATIVE. This kind of answer is deictic, i.e., the sememe—e.g., expressing location, person, fact, and so on—can be individuated with precision. In this sense, we say that *th-* is a constructive existential correlative answer ($\overline{\exists}$). By contrast, there are non-constructive existential answers ($\exists$) in English, like *some* or *many*. In English, uncountable nouns refer to a class of elements whose existence is non-constructive: for instance, the word *people* refers to a class of persons, but the individuals that form *people* cannot be found with precision, i.e., it is impossible to name an individual of the class.

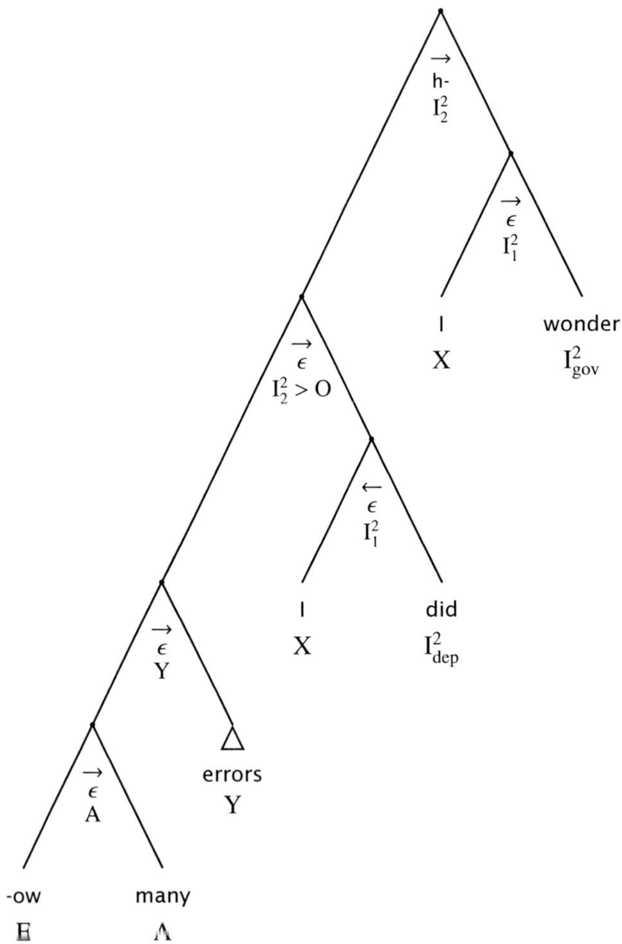

Figure 4.44: The adtree of *I wonder how many errors I did*

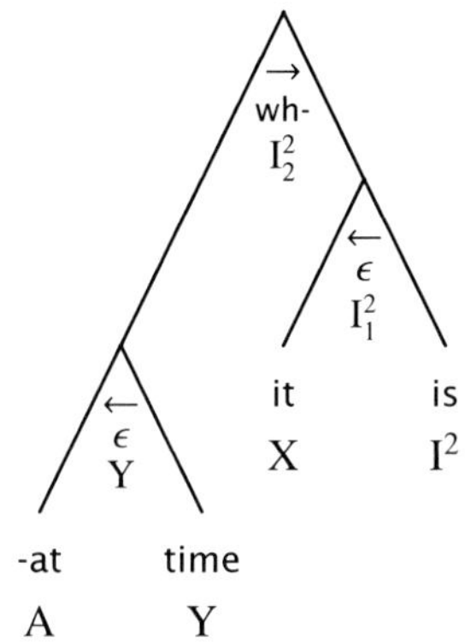

Figure 4.45: The adtree of *What time is it?*

Table 4.3: Constructive existential correlative answers in various languages

Language	*persona*	*factum*	*causa*	*locus*	*tempus*	*modus*
English	**th**-at	**th**-at	because	**th**-ere	**th**-en	so
German	**d**-er	**d**-as	**d**-arum	**d**-a	**d**-ann	so
Catalan	**t**-al	això	per això	allà	llavors	així
Polish	**t**-amten	**t**-amto	dla**t**-ego	**t**-am	w**t**-edy	w**t**-en sposób
Japanese	**ko**-no	**ko**-re	motte	(a)ko-**so**	sate	**ko** -chira

4.7 Causatives and factitives

Until now, we have put together two or more phrases, where part of the predication got hidden by construction, but we never dealt with explicit valency value augmentation. In this section we deal with some common constructions that increase the valency value of the verbant, with a special attention to what happens to the involved actants.

A standard construction that increase the valency value is the causative, used to introduce an extra in-valent actant, the CAUSE (animate or not), who cause the standard action to be done. Sometimes causative are completely grammaticalized, that is, the verb itself contains the sememe CAUSE. For example, the English verb *to kill* is the causative of the verb *to die*. In other cases, there are less grammaticalized constructions, where the standard action is still recognisable within the causative construction itself.

Let Q be the extra valent which brings the sememe CAUSE. We will show how the causative construction increases the number of the whole set of in-valent actants by one.

	$\overrightarrow{X}$		$\overleftarrow{Y}$
(4-40a.)	Karl	giest	die Blumen.
	Charles	*water*-3rdPERS	*the flowers*

	$\overleftarrow{Q}$		$\overrightarrow{X}$	$\overleftarrow{Y}$	
(4-40b.)	Ich	lasse	Karl	die Blumen	gießen.
	I	*make*	*Charles*	*the flowers*	*water*-INF

	$\overleftarrow{Q}$		$\overleftarrow{Y}$		($\overrightarrow{X}$)
(4-40c.)	Ich	lasse	die Blumen	gießen	(von Karl).
	I	*make*	*the flowers*	*water*INF	*(by Charles)*

Example (4-40a) shows the standard action— note that *Blumen gießen* is an idiomatic expression in German.

The causative construction (4-40b) is shown in Figure 4.46, where the infinitive became exceptionally an adposition, while the extra actant Q let the standard action happen: therefore, its trajectory is of government (←).

In (4-40c) there is the so-called 'causative passive', where the original doer is omitted, as in the passive, but it can be recovered as an external argument (X). [23]

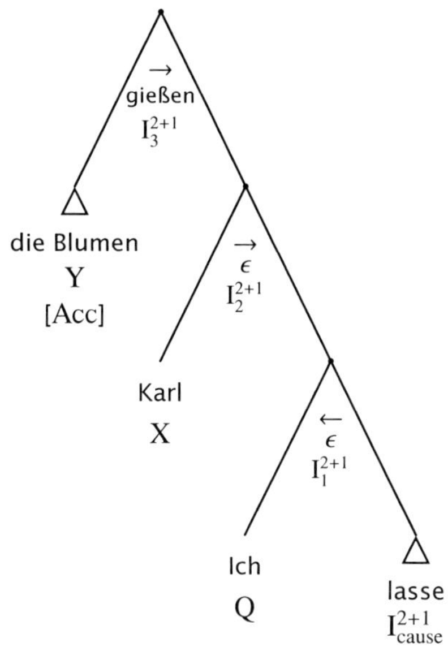

Figure 4.46: The adtree of *Ich lasse Karl die Blumen gießen*

The most extreme valency augmentation happens with quadrivalent verbs (I_{cause}^{4+1}), as in the following example (Figure 4.47).

(4-41.)	$\overrightarrow{Q}$		$\overrightarrow{X}$		$\overleftarrow{Y}$	$\overrightarrow{Z}$	$\overleftarrow{W}$
	Paul	makes	Liza	buy	a car	by Ron's	for 8,000 pounds.

Incredulity constructions as hidden causatives

There is a particular case of valency augmentation in children speech, called 'incredulity construction' by (Tomasello, 2003).

(4-42a.) him be a doctor!

(4-42b.) my mother ride a train!

(4-42c.) her wash the dishes!

Our analysis shows that an actant like *him* in (4-42a) can be considered a "subject in the accusative case" (Tomasello, 2003, sic, 103–104), but perhaps

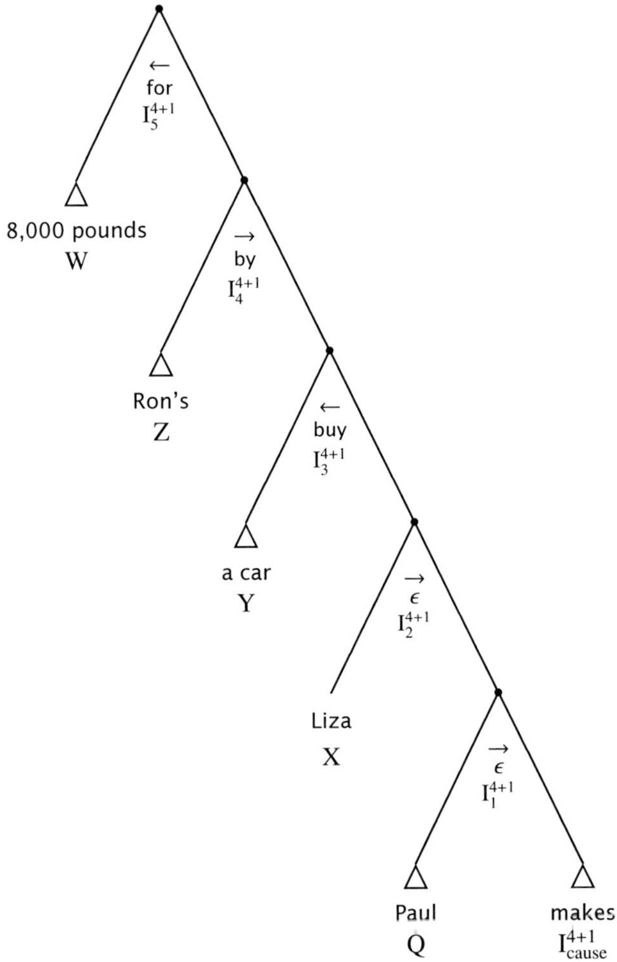

Figure 4.47: The adtree of *Paul makes Liza buy a car by Ron's for 8,000 pounds*

it is better to see incredulity constructions as relics of causative constructions, governed by a completely hidden *make/let/help*, as in (4-43), see Figure 4.48.

(4-43a.) (let) her do it!

(4-43b.) (help) him put his shoes!

(4-43c.) (make) this stay open!

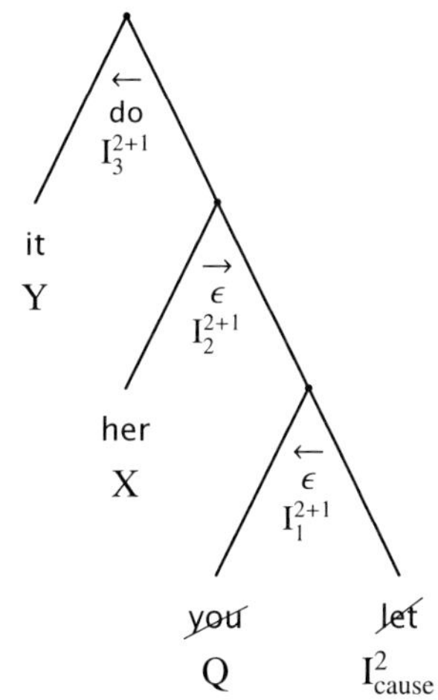

Figure 4.48: The adtree of *let her do it*

Chapter Five

Constructive dictionaries

Linguistic knowledge is a whole; while a strict separation of autonomous levels should not be considered—a tenet followed by cognitive linguists, e.g. Langacker (1990)—nonetheless it is useful to analyse languages on different levels, from sound to thought: phonology, morphosyntax, semantics, and pragmatics. In the previous chapters we have treated morphosyntax in details, while semantics played a minor role. [1]

We distinguish encyclopædic *vs.* dictionary knowledge within a language: meaning cannot be reduced completely to use—unlike what argue the followers of the so-called 'pan-encyclopædic' approach to language—because languages show morphosyntactic regularities that influence semantics, and this is the reason why semantics can and must be distinguished from pragmatics. [2]

We have previously seen that a phrasal construction is activated by the verbant and the actants put into relation through adpositions: the phrasal construction gives us a meaning that does not result *only* from the sum of the semantic traits belonging to the adpositional tree leaves, but also and mainly from the configuration patterns of the morphemes they belong to—i.e., the construction. [3]

In other words, the meaning of constructions does not result from the sum of their parts, i.e., the whole is more than the parts. On the other hand, it is also true that semantics is compositional, and thus dictionaries can be built accordingly. [4]

Constructive dictionaries are devoted to explain how morphemes are related within every language, i.e., how they are put together in order to have place in the constructions valid in that specific natural language. [5] As a direct consequence, each dictionary is language-dependent. Nevertheless, as the abstract rules to form adpositional trees are not language bound, we can sketch general principles to design constructive dictionaries, principles valid regardless of the language. How is this inventory of meaning structured?

As already shown in Section 4.3, in our perspective, a morpheme is formed by the morph, the signifier (explicit, phonetic sign in the syntagmatic axis) and one or more sememes, i.e. semantic traits.

Two morphemes which have different morphs but same sememes are called *allomorphs*: for instance, the English morphemes *-hood* and *-ship* are both suffixes used to denote the time of being of what denoted by the suffixed noun, as in the following pairs: *boy* ⇝ *boy-hood*; *dictator* ⇝ *dictator-ship* (examples from (Stein, 2007, 73,152–153)).

Vice versa, two morphemes which have the same morph but different sememes are called *homonyms*: the English morph 'bachelor' can be referred to at least four different morphemes: $bachelor_1$, 'an unmarried adult man'; $bachelor_2$, 'a male bird or mammal without a mate'; $bachelor_3$, 'a person who has taken a first university degree'; $bachelor_4$, 'a knight serving under the banner of another knight'. Each of these morphemes in fact is defined by a different configuration of sememes, and consequently they are used within different constructions: it seems perfectly reasonable to intend them as autonomous systems. [6]

The configuration of sememes belonging to a particular morpheme can be represented as an ontology: each sememe is collocated at a level of abstraction, with one parent sememe and zero or more children sememes. In other words, there is a relation of hyper- or hyponymy within sememes. Each sememe can have a list of hyponymies, such as: [VERB: [TENSE: [PERSON: [3RD, SINGULAR]]]]. Conversely, each sememe has at most one parent. For instance, the sememe TENSE is hyponymy of VERB, while 3RD and SINGULAR are children of PERSON. The whole family belonging to PERSON is hyponymy of TENSE.

It is obvious that sememes and their configurations are strictly language-bound: in Lihir, an Austronesian language spoken in the New Ireland Province of Papua New Guinea, there is a rich distinction in the hyponymies of the sememe NUMBER: beyond SINGULAR and PLURAL, it has the following hyponyms: DUAL, TRIAL, PAUCAL for two, three, few referents, otherwise the generic PLURAL is used (Corbett, 2000, 25).

It simply has no sense to use the sememe PAUCAL in the English constructive dictionary, because in English this sememe does not exist—meaning that such sememe does not affect the language constructions. [7] This is the reason why we need to build different constructive dictionaries according to the languages.

Ordinary dictionaries are cemeteries of words: a definition is given, along with some examples, from which the reader should evince the allowed constructions and how to use them. Constructive dictionaries aim to be different: the centre of the dictionary is the set of lexemes belonging to that language, along with the constructions that transfer their main grammar character in the other ones. What we will see in the next section is precisely *transference*, this process of grammar character change of lexemes. At the same time, we will define lexemes among morphemes and therefore what we mean by lexicon.

5.1 Transference

Every language, regardless of the Sprachbund it belongs to, has a class apart within its morphemes devoted to convey the most part of semantics. In our perspective, semantics is represented through the relations between sememes and also through the relations from sememes to the non-linguistic world, that is the reference. Morphemes belonging to the class referring the world are called *lexemes*. The class containing all the lexemes of language is called the *lexicon*.

While the concept of 'word' is central only in the grammar traditions of the Standard Average European, the distinction of lexemes within a language's morpheme set is general. For example, in Japanese non-lexemes are written in *kana* (syllabic script), while lexemes are written in *kanji* (Chinese logographic characters) (Yamasaki, 2000, 75). So, in the following, we will use the term 'word' in an oblique sense, meaning the apparently basic textual unit of a language, which corresponds to the usual notion of word in the Standard Average European Sprachbund. As already explained in Chapter 3, the basic unit of a language in the adpositional approach is the morpheme and a word is a linguistic structure composed by morphemes via appropriate constructions—and these constructions are the subject of this section.

Normally, the word 'lexeme' means the basic lexical unit, that is, the word part—sometimes called the 'root'—where semantics is mostly conveyed.

Within a single language, word classes forming a fixed set, such as pronouns, determiners, prepositions, etc., are usually non-lexical: in some grammar traditions they are called 'functional words' or 'formal words' [8]. On the contrary, classes that belong to arbitrary sets, e.g. common nouns, verb roots, etc., are lexemes. In general, lexemes show a lesser degree of grammaticalization compared to established, fixed word classes like 'functional words'. It is easy to show this difference through a contrastive example.

(5-1a.) Nel mezzo del cammin di nostra vita, mi ritrovai per una selva oscura.

(5-1b.) Mezz' cammin nostr' vit', 'trov' selv' oscur'.

(5-1c.) Nel -o del di -a -a, mi ri- -ai per una -a -a.

Example (5-1a) is the first phrase of Dante's *Divina Commedia*: the quotation is still recognisable to anybody who knows it in advance when non-lexical morphemes are cut off (5-1b), while the reverse—cutting off lexemes—is not true (5-2c). Therefore, lexemes convey the most part of meaning, i.e., they have more sememes than non-lexical morphemes.

But the dichotomy 'fixed *vs.* open sets of word classes' is unsatisfying within a constructive perspective [9]: there are some constructions, partially grammaticalized, that can be put at work only by a fixed set of morphemes and therefore this set forms a fixed word class. Therefore, we need more solid criteria to distinguish the open set of lexemes from the generic set of morphemes.

The criteria we need are the grammar characters. It was Tesnière the first to use grammar characters to find the inner structure of syntax, in the 1930s. In particular, he found that there are some morphological and syntactic strategies whose main purpose is to 'turn' the phrasal or sub-phrasal group of morphemes. This phenomenon was called in French *translation*, while in English "an equivalent may be *transference*, as the word *translation* has already the meaning of the French 'traduction'." (Tesnière, 1959, 367, our translation). [10]

From a formal point of view, a *transfer* (a single act of transference) is only an adtree transformation. However, it has a relevant linguistic meaning: most transfers are so acclimated to speakers by grammaticalization that they are no longer aware of their existence.

Transference can be stative, adjunctive, verbant or circumstantial, following the grammar character of the Tesnerian *transférende* (literally, 'transfer-worth'), i.e., the morpheme where to start the transfer chain, which possesses a proper grammar character. What Tesnière called a *transférende*, for us is simply a lexeme. In other words, a lexeme is a morpheme which carries a grammar character of its own, and so it can stand as the governor in an adpositional tree (Figure 5.1).

Conversely, non-lexical morphemes always performs transference starting from a lexeme. At the limit, the transfer can be redundant (e.g., O > O).

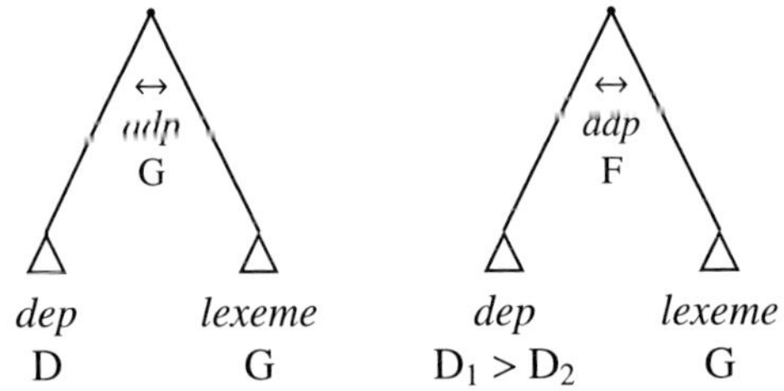

Figure 5.1: The place of lexemes in the abstract adpositional tree structure

Of course, it will be more common to have verbant and stative lexemes than adjunctive or circumstantial ones: however, this is not a rule valid in general, only a natural tendency of languages.

From a purely combinatoric point of view, each type of transference can only have three possibilities. For example, there are three possible stative transfers: to verbant, to adjunctive, to circumstantial. It is worth noticing that sometimes grammaticalization prevents a specific type of direct transfer. For example, sometimes in English to pass from stative to circumstantial (O > E) morphology requires to pass through another type of transfer, the adjunctive (e.g., O > A > E). The word *histor-ic-al-ly* is a good example: it is a circumstantial transfered from *two* adjunctives—*histor-ic-al* and *histor-ic*, which are not the same—while the lexeme is the stative *history*. There is no morphological shortcut to obtain the circumstantial: **historicly* and **historly* are not acceptable.

(5-2a.) essen > das Essen (I > O)

(5-2b.) to eat > an eating (I > O)

(5-3a.) acquist-are > acquist-o (I > O)

(5-3b.) to buy > a buy/a bought (I > O)

Examples (5-2) and (5-3) show two verbant-to-stative transfers, in German and Italian, and their correspondent transfers in English. The difference between German, English and Italian in this case is the following: Italian shows a higher degree of grammaticalization, as it prefers a morphological strategy, while German and English show a lower degree, preferring a syntactic one.

(5-4a.) ricco/a > i ricchi (A > O + [-DEFINITE, COLLECTIVE, PLURAL])

(5-4b.) rich > the rich (A > O + [-DEFINITE, COLLECTIVE, PLURAL])

Sometimes a transfer is marked, i.e., the grammar character of the transferred word (i.e., the result at the end of a transfer, the Tesnerian *transféré* [11]) has one or more sememes that mark the transfer itself, as in (5-4ab). Such *marked transfers* are adtree transformations where the lexeme can belong only to a limited list, with precise features that bound the application of the transformation.

Figure 5.2 shows the abstract constructions behind (5-4a). The constructive dictionary of the Italian grammar will have a section devoted to (adjunctive) lexemes, such as the governor of (5-4a), and a section devoted to abstract

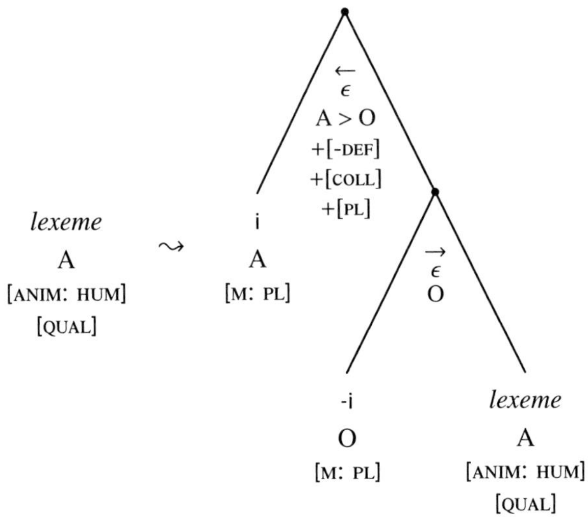

Figure 5.2: The construction of the adjective-to-stative transfer *i ricchi*

adtrees—i.e., constructions [12]. If the adjunctive lexeme has the proper sememes, in the present case, ANIMATE: HUMAN and QUALITY, then this construction is admissible. The construction itself imposes some sememes: the masculine gender (M) and the plural (PL) in the leaves. The plural also labels the whole group, along with the sememe COLLECTIVE and the marked *lack* of DEFINITION, i.e., the fact that the members of the group (the rich, the poor, and so on) cannot be individuated, or, if you prefer, in mathematical terms, the set denoted by the adtree is non-constructive, that is, it is not possible, a fortiori, to indicate an element in the set in a canonical way.

In general, grammar characters and the semantic configuration of a lexeme, i.e., the required sememes, determine the conditions enabling the application of a construction on having that lexeme as a governor.

5.2 Strategies of grammar character change

Whorf (1945) distinguished lexemes in two categories: lexemes by selection and lexemes by collocations. Lexemes by selections are all those lexemes whose grammar character is known at a first glance: among his examples in English, he gave *police* (O), *elephant* (O), *honest* (A).

By contrast, lexemes by collocation are defined only if put into the syntagmatic axis: we can have cues about their most probable grammar character, but we cannot be certain. For instance, the English lexeme *walk* is probably a verbant (I), but it can also be a stative (O) or even a adjunctive (A):

(5-5a.) I **walk** in the park (verbant collocation).

(5-5b.) Let's have a **walk** (stative collocation).

(5-5c.) **Walk**ing distance (adjunctive collocation).

In (5-5), the proper grammar character of *walk* is defined by the constructions where it is collocated, more specifically:

(5-5a.) O + *walk*

(5-5b.) (A + DETERMINER) + *walk*

(5-5c.) *walk-ing* + O

While selection is a property of certain lexemes, collocation is a *strategy* used to change the grammar character of lexemes—a strategy based on syntax. Of course, selective lexemes can change their grammar character by collocation. For instance, in *elephant ivory* the stative lexeme *elephant* was turned into an adjunctive (O > A) by collocation.

Apart from collocation, there are only other three possible strategies to modify the grammar character of lexemes within word morphology, used alone or together: derivation, ablaut, and suppletion.

Derivation

In order to change the grammar character of lexemes, most languages have morphemes devoted to this task, that are morphologically applied to lexemes. This kind of transference is usually called *derivation*: we retain the common use because there is no ambiguity. In fact, the result of a derivation is always a word.

Sometimes derivation is performed not to lexemes but to special classes of words, treated lexically. For example, the English morpheme *-fold* is a number-adjectives transferrer: it is applied to words denoting cardinal numbers, treated as lexemes [13], in order to form adjectives that show the amount of the number specified, having as many parts as stated in the cardinal number: *threefold*, *two-hundredfold* (Stein, 2007, 58).

If a morph can perform different derivations, i.e., involving different grammar character changes or marking the transfer with different sememes, it will be treated as belonging to different homonymic morphemes. For example, there are many English morphemes *-s* with very different values (Stein, 2007, 149):

(5-6.) (the) elephant**s** (O > (O + PLURAL)).

(5-7.) (he/she) beg**s** (I > (I + 3^{rd}PERSON: SINGULAR)).

(5-8.) (the) gape**s** ((I + HUMAN: STATE) > (O + PLURAL, HUMAN: STATE: QUALITY)).

(5-9.) (on) Monday**s** ((O + TIME) > (E + TIME: DURATIVE)).

The example(5-6) shows the most common usage of the morph '-s' in the English language: to pluralise nouns. Instead, (5-7) shows another common construction in English: to mark the third person singular of verbs.

The example (5-8) shows a more limited construction, as the verb should describe a state, condition or disease, made stative by the construction, e.g., *the jitters*, *the strangles*. In other words, the lexeme should be a VERB with at least the following sememes: HUMAN: STATE. The construction itself leads as a result a stative group of morphemes, having the configuration shown in (5-8).

Finally, the derivation shown in (5-9) shows a durative construction, which can be applied only to stative lexemes having TIME as the core sememe.

Ablaut

Some languages—the German ones in particular, but not only—can 'rotate vowels' in order to change grammar character or to add or change particular sememes. For example, in English there is a Germanic verbal paradigm, i.e., a particular verbant construction, that changes the present tense in the past tense and in the past perfect, as shown in Table 5.1. The transfer here is verbant-to-verbant (I > I): the relevant information is the change of the hyponym of TENSE.

In Hebrew, derivation is performed through given vowel patterns, while lexemes are precise sequences of consonants. The principle is the same as before, i.e., ablaut (Table 5.2).

Table 5.1: Germanic verbant construction in English performed by ablaut

	present	*past*	*perfect*
i:a:u	sing	sang	sung
i:a:u	drink	drank	drunk
i:a:u	begin	began	begun

Table 5.2: Stative (*e-e*) and verbant (*ā-a*) transfers in Hebrew (example)

	e-e		*ā-a*	
b'r'k	*berek*	'knee'	*bārak*	'he kneeled'
d'r'k	*derek*	'road'	*dārak*	'he marched'
g'b'r	*geber*	'strong man'	*gābar*	'he was strong'

In such cases it could be difficult to define which is the default grammar character: an indication is to take the grammar character of the citation form. For example, for *b'r'k* the citation form is usually *barak*, which is a verbant meaning 'to kneel, to bless'.

Suppletion

The grammar of languages is the result of grammaticalization. This means that the most frequent patterns of usage gives raise to productive constructions, and they form what speakers call 'rules' of the language. Grammaticalization is a complex driving force, which is partially guided by biological limits, partially by language policies and planning actions (or omissions) made on the language itself by humans. Language policy and planning in constructive linguistics is, by the moment, a further work, and hence it will not be treated in this book.

However, what is important to underline here is the fact that sometimes grammaticalization fails in its goal; after all, languages are made by humans, and humans are far from being perfect. This is the reason why ad hoc constructions—quite often lexically bound—used to perform a single transference, are not rare in natural languages. [14]

(5-10.) mouse > mice (O > (O + PLURAL)).

(5-11.) tooth > dentist ((O + OBJECT: TOPIC) > (O + TOPIC: HUMAN: PROFESSION)).

(5-12.) go > went > gone.

For instance, the English plural of *mouse* is not **mouses*—save perhaps some variants of children language. In fact, the construction exemplified in (5-6) is not active, and an ad hoc lexeme is used: *mice*. Suppletive strategies are expensive, both for the language learner, and for the constructive dictionary, as there is no shortcut: they should be coded inside by hand.

Analogously, the transfer from *tooth* to *dentist* is suppletive, as the most used derivation strategy of English would require **toothist*, as in the case of *journal*, from where *journalist* is derived.

Many languages having a long and important cultural tradition, as English, show suppletion strategies in the original foundation of their lexicon. (5-12) shows that the present and perfect forms of *go* derive from Old English *gān*, "related to Dutch *gaan* and German *gehen*" (Oxford Dictionary), while the past form derives from the past form of another Germanic verb, *wend*, which supplied the lack of past form in *go*.

Every time a language is forced to use unrelated lexical material in order to perform transference, it works by *suppletion* and the unrelated material is called 'suppletive' because it supplies a failure of grammaticalization. Luckily, there is no natural language completely suppletive, as grammaticalization does work well most of the times. [15]

5.3 Toward constructive dictionaries

A reasonable constructive dictionary of a language would require an entire volume, not just a single section. Furthermore, it is simply impossible to write down a *complete* dictionary of a living language—constructive or not—because any corpus is only a snapshot of a partial language-in-use and in a given moment: the moment after the release of the corpus itself it starts to get old.

Moreover, each individual's repertoire is unique. In general, most of the lexicon, the morphemes and the constructions are in common with other members of the same speech community, but every speaker has a proper language style, uses constructions in a slightly different manner, etc. It is what linguists usually call 'idiolect': every dictionary is bound to be incomplete.

These are the theoretical linguistic limits. Our aim here is foundational, i.e., we want to give the right hints to build a constructive dictionary. From a theoretical point of view, there are at least two parts in a constructive dictionary: morphemes and constructions.

First, a constructive dictionary should list morphemes and their properties. As we have previously seen, morphemes are defined by: a morph (possibly empty); a grammar character (for lexemes) or a transfer (for non-lexemes) [16]; and, finally, a configuration of semantic traits—its sememes (organised along hyper- and hyponymy).

Second, there is a list of constructions, i.e., abstract, non-instantiated adtrees where to place morphemes, following the appropriate structural requirements, i.e., the right grammar character at the right place in the adtree, the right trajectory of information prominence, and so on. The number of constructions could be very high, but this number can be reduced thanks to the fact that most constructions can be described in terms of transformations, starting from verbant constructions where valency is saturated—see again Chapter 4 if needed.

From the description given above, it should be clear by now that a constructive dictionary describes far more than a dictionary in the usual sense—i.e., a book that lists the words of a language in alphabetical order along with their meaning—but rather the grammar of the language in a broad sense: morphology, syntax, lexicon, and semantics—only phonology and phonetics are excluded.

Nevertheless, we start our analysis to propose an organisation for a constructive dictionary by making a step back, and looking more carefully to traditional dictionaries, to understand how they fail to adapt to a constructive approach.

Traditional dictionaries in a constructive world

A usual dictionary is an ordered list of words, each one with its attributes. In a normal dictionary, a word like ‘walk’ can be listed as a noun or a verb, transitive or intransitive, according to the meaning. Also, each entry for ‘walk’ uniquely describes a possible meaning, along with a few examples. Finally, some common patterns of usage are listed as well, for example ‘walk on air’, ‘walk off’, etc. Some forms, like ‘walks’ or ‘walking’ are not listed since they are canonical modifications with a meaning that can be immediately inferred, see example (4-1) in Chapter 4.

From a constructive perspective, we see that ‘walk’ is a morph, having a number of homonyms, one for each entry in the dictionary—where an entry describes a single morpheme. Moreover, each morpheme has a number of attributes, either structural—e.g., being a noun or a verb, transitive or intransitive, etc.—or semantical, as the description really identifies the semantical traits that characterise the morpheme, although not in a formal way. The common patterns of usage are nothing else than grammaticalized constructions, which have become so 'hard-wired' to gain an outstanding status and meaning, so they are reported in the dictionary. Some of these construction are mostly idiomatic, like 'walk on air', and some are still productive, like 'walk off'.

In a constructive dictionary, there are no words, as the basic units are the morphemes. Mimicking the traditional approach, each morpheme appears along with its attributes and meaning. The attributes are the grammar character or the transfer, the fact of being unergative or unaccusative if it acts as a verbant, attributive or predicative if it is an adjunctive, the sememes, etc. All these attributes have been discussed in the previous sections and chapters of this book.

Obviously, we distinguish between morphemes having the same appearance, i.e., the same morph, but different attributes or sememes, which, in turn, are a special kind of attribute. These morphemes are grouped together as homonyms, as words with the same writing but different meanings are grouped together in a traditional dictionary. Also, we use sememes as a way to describe the meaning of a morpheme, and the structural attributes are the proper ones of the adpositional paradigm, that is, the grammar character instead of noun, verb, adjective, and so on. But we retain that each entry is a unit describing exactly one morpheme with its attributes, as each entry in a traditional dictionary describes a single word in a specific meaning and usage.

In this way, we obtain a constructive dictionary that can be coupled with a constructive grammar, i.e., a collection of admissible constructions and a set of validation rules, which are properly called ρ-transformations, that discriminate valid results of the application of constructions from the just structurally-sound ones, which did not take their place in the language.

Unfortunately, this picture is largely inadequate and unsatisfactory. In fact, constructions in the adpositional paradigm are abstract entities, see valency in Section 2.3, depending on their governor for a correct instance. Recalling example (2-3) in Chapter 2, although it is perfectly correct to regard the morpheme 'open' as a trivalent verbant, the application of the valency construction is not immediate, as each actant is linked to the governor via an adposition. And different trivalent verbants, although enabling the same construction, require different adpositions to introduce their instances of the three actants.

Hence, as valency is a fundamental construction in every language, it cannot be completely separated from the nature of the governor. Or better, we may divide governors into classes: one class having each actant introduced by an empty adposition; one class where the first and the second actant are introduced by an ϵ and the third one by 'to'; another class for a different combination of adpositions, and so on. The price to pay for operating such a partition is to definitely obscure the nature of valency as a unifying concept, shared by each natural language. This effect is unacceptable for a constructive point of view, as ours, since it really forgets the very construction, keeping just many possible instances, instead of focusing on the fundamental aspect, valency in this case, hiding the details, as we silently did most of the times in the previous chapters.

Hiding is a nice principle since it does not lose the details, the adpositions introducing the various actants in the example, but simply it allows us to recover them on need, or to ignore them when disturbing. In other words, hiding allows recovering, while what is forgot, is lost.

Thus, we are saying that a strict separation of the grammar, intended as the rules to build the language, from the dictionary, intended as the set of the admissible morphemes, is artificial and, ultimately, counter-productive since it forces to distinguish and to separate what is unifying in the construction of expressions.

In this sense, the example of the valency construction is very significant, as valency, as already remarked in Chapter 3, is the prototypical construction, and any other construction, in a very abstract sense, can be seen as an instance of it. This has been formalised in the notion of *grammar product*, introduced in the notes on Chapters 1 and 2, and explained in Appendix B, a fundamental piece of the formal model.

A mathematically inspired dictionary

As evident from the previous discussion, a constructive adpositional grammar for a natural language cannot separate morphemes from constructions. So, a dictionary should contain both of them, and should describe them as coordinated elements, allowing the construction of all the well-formed expressions of the language. As already stated in Section 4.3, this is not possible without considering also redundancy, which is encoded by ρ-transformations.

In principle, these ingredients—morphemes, constructions and ρ-transformations—suffice to describe and to generate the whole language. So, a dictionary can be abstractly described as their collection along with an organisation principle allowing for their inspection on need.

In fact, we can conceive the collection of our ingredients as a *database*, no matter how internally represented—after all, computer scientists have developed a series of very sophisticated techniques to solve the representation problem—that can be accessed via a limited number of 'queries' supported by an appropriate indexing system. What really matters is to define what we want to index, as it determines what we can search for.

In this sense, morphemes are easy: each morpheme is described by a morph and a set of attributes. It is reasonable to search for a morph, retrieving the set of corresponding homonyms, or to search for a specific attribute, retrieving the set of morphemes having it in their representation. [17] Evidently, in this view, a morpheme is a record composed by a morph and the list of its attributes.

Also, the representation [18] of morphemes is natural and coherent with the treatment we made of them, as depicted in the previous chapters. To cope with suppletion it is useful to have 'cross-references', that is, links between pairs of entries, marking the special relations of the couple of morphemes. In fact, these relations can be formally represented by transformations mapping a root-only adtree into another root-only adtree.

We notice that the collection of morphemes is potentially open on two levels: the collection of morphemes is not assumed to be exhaustive, i.e., there may be morphemes which are admissible in the language but not known to the dictionary compiler, and so they have not been included; also, the list of attributes of a single morpheme may not comprehend all the possible attributes, in particular sememes, as a particular 'meaning' of a morpheme is not in the domain of the dictionary compiler.

A slightly more complex situation is that of constructions: in fact, there is a minor ambiguity in the use we made of this word, and now it is time to clarify. In fact, we used the word 'construction' to identify three distinct entities.

In the first sense, a construction identifies a family of adtrees where the governor and the adpositions are fixed, and the tree structure is rigid. What may vary are the dependents, that should match the specified attributes, primarily grammar characters. We will name these forms as *basic constructions*. An example is (2-3a), whose adtree is shown in Figure 2.12, 'X opens Y with Z'. Here, the governor 'opens' and the adpositions ϵ, ϵ and 'with', introducing the actants X, Y and Z respectively, are fixed while the actants may be any adtree one likes, provided it is a stative, i.e., it has the O grammar character.

As a basic construction is an adtree lacking some leaves, it can be easily represented in the database by a logical term with variables. For example, the term representing the basic construction of (2-3a) is with(ϵ(ϵ(opens, X), Y), Z), i.e., the linearised adtree where the actants are substituted with the X, Y and Z variables and the adpositions are function symbols.

In the second sense, a construction identifies a family of adtrees whose governor and dependents may vary, although they have specific grammar characters (or specific attributes), but the adpositions and the tree structure are fixed. We call *general constructions* these forms. It is obvious that basic constructions are just partial instances of general construction, and so any ambiguity is immediately solved. An example of general construction, where the adposition has a fixed value, is shown in Figure 3.24, which illustrates the 'generic' circumstantial-verbant relation.

Evidently, general constructions can be represented in the same way as basic constructions, using logical terms where no lexeme is present. An important point to remark is that logical terms are not linearised adtree, although they share the same graphical representation. In fact, the logical terms are really functions, parametrised by their variables; only when those variables get a value, the result becomes a real adtree.

Being representable as logical terms, general constructions can be suitably represented and indexed in the database by standard techniques of computer science. In particular, searching is performed via pattern matching: given a logical term Q, the query, we may ask for all the (general) constructions whose representations are instances of some subterm of Q. The term matching problem has been deeply analysed in literature, see, e.g., Kilpeläinen (1992), for example in the implementation of the Prolog language, see Baader and Siekmann (1994).

It is reasonable to be curious about the relation of the mathematical model in Appendix B with the proposed representation of the dictionary; in this respect, it is worth noticing that what have been called 'general constructions' are exactly the same thing as 'grammar constructions' in the formal model—we prefer to retain different names to distinguish the contexts. Also, morphemes are just the atoms of the formal model, and ρ-transformations are nothing else than redundancy transformations. So we have a strict correspondence, essentially an identification, among our fundamental concepts in the linguistic, mathematical and database-oriented contexts.

In the third sense, a construction is a family of adtrees whose common feature is a sort of similarity among them, usually derived from a common usage or form. A typical example is the passive construction of bivalent verbants, analysed in Chapter 4. Our proposal is to consider those constructions, named *abstract constructions*, as families of general constructions.

After an extensive analysis, we have not been able to find any abstract construction which naturally occurs in the study of linguistics and that cannot be reduced to a (finite) family of general constructions, so we induce that defining abstract constructions precisely as finite families of general constructions is a good definition which covers the general case.

Being a family of general constructions, an abstract construction α can be immediately represented in the database as a query, specifically as the logical term $\epsilon(C_1, \ldots, C_n)$ where C_1, ..., C_n are the terms representing the general constructions which are members of α.[19] In fact, pattern matching will produce as a result exactly the instances of C_1, ..., C_n in the database, thus reducing the problem of representing abstract constructions to the already solved problem of representing and searching for general constructions.

In the same line, one can easily represent ρ-transformations. In fact, a ρ-transformation, see Section 4.3 if needed, maps adtrees matching a fixed pattern into adtrees having a similar structure, the idea being that the pattern individuates the 'mistakes' and the result is a 'corrected' adtree; the parametric parts of the pattern are used to instantiate the result. If we represent a ρ-transformation as a pair of logical terms, the first one for the pattern adtree, the second one for the resulting adtree—with the additional care to use the same variables where needed—the application of a ρ-transformation to an adtree A can be described as follows: first, we check that A matches the pattern of the transformation—this operation is just matching two logical terms; if the match is possible, we will have an assignment of the variables in the pattern making it equal to A; then, the assignment is applied to the result part of the transformation, yielding the B adtree, which is the result of the application of the ρ-transformation to A. Please, compare this procedure with examples (4-1) and (4-2) in Chapter 4. Please, notice that also transfers can be represented along the same line: see Figure 5.2.

Since a ρ-transformations can be represented as a pair of terms (P, R), it can be more efficiently represented by the single term $\epsilon(P, R)$, which suits the format of our dictionary as designed so far. And, of course, the query mechanism based on term matching, and used for searching among constructions works also to search for ρ-transformations.

Hence, using a database as our dictionary, where morphemes are represented as records composed by the morph and the attributes, and constructions and ρ-transformations are represented as logical terms according to the linearised form of the involved adtrees, provides us with precisely the organisation we want for a constructive dictionary, overcoming the difficulties we have found in a more traditional approach. The key aspect behind this organisation is that matching of logical terms is computable and reasonably efficient in practice. Also, the morphemic level of the language can be represented, since transfer share the same representation and searching strategies as ρ-transformations.

One may be tempted to slightly extend the definition of general constructions, allowing for variable adpositions. And the reader may wonder why we have not proposed such an obvious generalisation in the first place, since it seems to reduce the number of constructions and ρ-transformations we may need to represent—see, e.g., Figure 3.24 representing the generic circumstantial verbant relation, now with an adposition which is not fixed.

The point is that matching is delicate: having variables for adpositions means to have variables for function symbols, so the problem becomes a second-order matching which, in general, is NP-hard, i.e., there exists no efficient algorithm which is able to compute a matching assignment, as shown, for example, in Hirata et al. (1999). So, our representation is exactly fit to the computational needs of computer-supported dictionary.

A sketch of a real-world constructive dictionary

The dictionary illustrated so far is perfectly adequate for a computer-based implementation, as a substitute for a traditional dictionary, especially for automated applications. Being an almost direct implementation of the formal model, the dictionary is also useful as a reference model for organising the information about the structure of a language, what is usually called its grammar.

Nevertheless, from many points of view, a different sort of dictionary is desired. In the first place, a dictionary should be agile: it should contain the information one needs and this has to be compact. On the other side, the organisation can and should be significant for the reader, enabling for a deeper understanding of the language.

Starting from the second remark, our reference dictionary is poor: it does not contain the links between adtrees which are evident for a normal speaker. For example, it is evident for any speaker how to convert a phrase from the active to the passive form, or how to shift time from the present tense to the future tense or to the past one.

But these relations among constructions are completely absent from our reference dictionary. One may think that these relations are not essential to generate the language, and it would be correct. But still, it is hard to deny that understanding a language means to master these transformations.

On the other side, the reference dictionary is bound to be huge: a natural language has hundreds of thousands morphemes, and general constructions are a combination of adpositions, valency and grammar characters with an evident combinatorial explosion—which can only grow when considering also semantical traits and the other attributes of morphemes. A rough estimate for a language like English is about a thousand of constructions.

These facts are not surprising as no one is unaware of the relations among morphemes and among constructions—so the 'really different' morphemes and constructions are far less. We spent the entire Chapter 4 to study the relations among constructions—they are exactly the τ-transformations—and the first part of this chapter to study the relations among morphemes—called transferences.

Being aware of transformations is a way to compress the amount of information in the reference dictionary. We do not need to have a separate construction for the past tense or the future tense when we know how to transform an adtree in the I grammar character and in the present tense into another adtree where verbants are in the past or in the future tense.[20] Nor we need to understand how to deal with the passive diathesis of a bivalent verbant when we know how to deal with the active one and we know how to transform an active adtree into a passive one. Since most τ-transformations are applied to large families of adtrees coming from different general constructions, we obtain a significant cut in the number of constructions we need.

The same holds for morphemes: transference allows to consider a single morpheme and the 'cloud' of morphemes it generates. As far as transformations are widely applied and of general validity, most of the dictionary can be reduced to a few thousand of basic morphemes and a number of fundamental constructions—their number is hard to estimate, but no more than a few hundreds of general constructions.

So a practical dictionary, which is also useful and significant to understand the nature of a language, must represent basic morphemes, the fundamental constructions, the ρ-transformations to check the validity of the generated expressions, and also a number of τ-transformations and transferences.

We already know how to represent and to index in a database the morphemes, the constructions and the ρ-transformations. So we have to analyse how to represent tree transformations and general transferences in a way that make them suitable for efficient searching.

As said at the beginning of Chapter 4, τ-transformations and transferences are, in fact, instances of tree transformations in general, operating at different levels in the description of the language. So, they will share the same representation, since the context they are applied within changes but not their nature.

In principle, a tree transformation, being just an endofunctor in the category of adtrees and their constructions, may map any construction into any other, as far as composition, i.e., applying two construction one after the other, is preserved and identity constructions too. But this level of generality is needed to have a formal model powerful enough to capture languages of any sort, in particular artificial ones, see Appendix B.

When we deal with natural languages, as the results and examples of Chapter 4 and Sections 5.1 and 5.2 have shown, transformations operate as identities except for a few constructions being mapped into different ones—those which are significant for the transformation, being its focus and target. [21]

So, the transformations which should be reported in a dictionary, have a limited format: they apply only to a few general constructions, mapping them into arbitrary, but homogeneous constructions; in all other cases, they map a construction, i.e., an adtree, into itself.

Thus, we can represent a transformation as a set of pairs composed by a pattern and a result, both being logical terms representing adtrees. Each pair represents the transformation as applied to one of the target general constructions, and it operates by matching the pattern against a given adtree, getting a matching assignment which is in turn used to instance the result.

In this way, a tree transformation's representation generalises the one of ρ-transformations. In fact, ρ-transformations—not surprisingly—can be seen as tree transformations whose representation is a singleton set. And, of course, indexing and searching for transformations is a simple and evident generalisation of the corresponding operations on ρ-transformations.

For example, the active-to-passive transformation for a bivalent verb can be coded as the set T of pairs (P_a, R_a) where P_a and R_a are as in Figure 5.3, with a ranging over the class of adpositions—compare with Figures 4.12 and 4.14—and X, Y and V are logical variables.

Of course, the depicted transformations has to be corrected by suitable ρ-transformations to put the 'esse' verbant in the correct tense—we have not indicated the sememes not to burden the already complex picture—and, in some cases, see, e.g., (4-5), Figure 4.14, to consider suppletions.

Summarising, the entities composing the dictionary, which is a database, are represented as in Table 5.3, and the searching and indexing operations are naturally reduced to pattern matching of (first-order) logical terms.

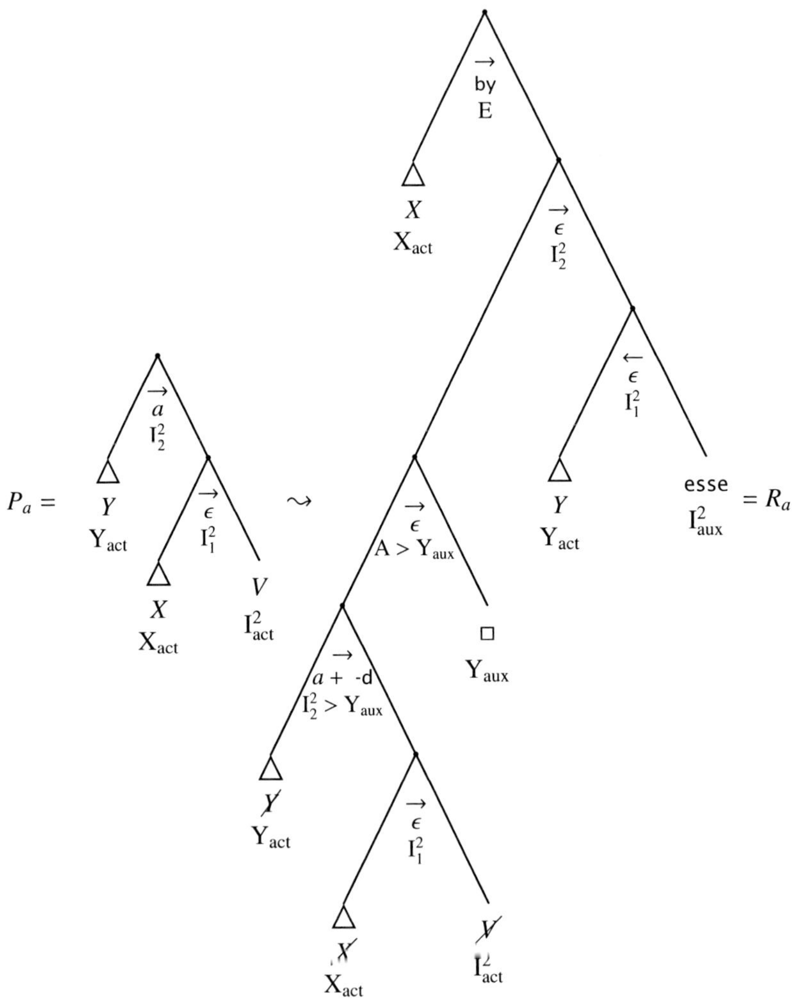

Figure 5.3: Representing tree transformations

Table 5.3: Representation of an adpositional grammar

Morpheme	M($m, a_1, \ldots, a_n$)	m = morph, $a_1, \ldots, a_n$ = attributes
Construction	C(c)	c = logical term corresponding to the linearised adtree
Transformation (ρ, τ and transfer)	T($\tau(p, r)$)	τ = name of the transformation, p = pattern, r = result; there are as many terms with the same τ as the number of components of the transformation

Chapter Six

Discourse and beyond

Why there is more than one single human language in the world if learning to speak a language is a natural fact carved in human beings? After dealing with morphosyntax and semantics we should finally put natural languages into use: this is the realm of pragmatics.

Usually, we have used the term 'language' referring to 'natural languages' taking this last expression as granted. However, this is far from being true: 'natural language', as well as other common expressions—such as 'mother tongue' and 'native speaker', for instance—are culturally and historically determined (Bonfiglio, 2010). Natural languages are conventions that refer to the social dimension of language competence and production, and every natural language emerges from grammaticalization, that is a cultural and historical process: there are many cultures in the world, and that is why there are many languages. [1]

Languages arose approximately 100,000 years ago by inter-subjective validity: unlike vocalisations and gestures of primates, our ancestors learnt by imitation, establishing social shared conventions that finally evolved in what we call nowadays natural languages. [2]

Unlike primate communication, human linguistic symbols are tools for directing attention. [3] In fact, linguistic symbols can be purely *declarative*, i.e. inform the receiver with no expectation of an overt behavioural response. In other words, a linguistic *act*—an actual utterance—construes a *scene*, i.e., something that goes beyond the straightforward perceptual and sensory-motor cognitive representation.

Speech act theory, pioneered by Austin, gives an explanation of this very important fundamental property of natural languages. When a single linguistic act is performed, each linguistic act says something (constative) and does something (performative) at the same time (Austin, 1962, 47). Within the adpositional paradigm, this 'doing things with words' is mapped into the adtree representation through the indicators of the information prominence trajectory. But adtrees only satisfy the need to represent what is actually said (the *locutionary* facet, in Austin's terms), not what is really meant, i.e., the *illocutionary* facet, nor what is actually, behaviourally, performed (the *perlocutionary* facet).

In other terms, each speech act has three facets, the locutionary one being properly linguistic in nature, while the non-locutionary ones—the illocutionary and the perlocutionary—are not entirely linguistic: they can be get explicit by linguistic means, but this cannot be given as granted.

In fact, what a speaker really means while performing a speech act is not always completely and explicitly clear in linguistic terms: nevertheless, at least in principle, the same speaker can express his meaning in explicit linguistic terms if required. The same can be said for behaviours activate by speech acts, even if the gap with the locutionary facet is even greater, as perlocution by definition is the collection of the *non*-linguistic responses to the speech act. For this reason, it seems not feasible to introduce perlocution into constructive grammars, which are *linguistic* models after all.

By this moment, our aim will be restricted to give an account of illocution, i.e., what the speaker means beyond the actual wording. Can the notation presented until here be extended to represent, if not perlocution, at least illocution?

The answer is positive: during the refinement of the linguistic and formal model of adpositional trees, our work raised the interest of the research group belonging to the European Center for Therapeutic Assessment at ASAG, Catholic University of Milan, Italy. [4]

The Center provides consultation, diagnosis and psychological treatment of family, relational and individual problems, following and developing a specific approach called Therapeutic Assessment (Finn, 2007). [5]

It is important to stress the fact that our application of the adpositional paradigm is a pragmatic analysis of linguistic data: there is neither clinical nor psychological evaluation. The linguistic analysis should be treated as a support for therapy, not part of it. Furthermore, even if it can be a useful support to Therapeutic Assessment, it is not specially linked to it.

The challenge put by the research group of the Center can be stated as follows: how can we extract pragmatic information from a text using constructive adpositional trees? In particular, how pragmatic information from the transcripts of conversation occurred in therapeutic settings can help therapist's and assessor's work? This chapter aims to answer these questions referring to the seminar mentioned above and the first analyses of transcripts already done.

This experience can be seen as the first real-world application of constructive adpositional grammars. The research program is to set up a framework for direct application of Constructive Conversational Analysis (CoCAL) suitable for therapists, in particular therapeutic assessors. [6]

6.1 How to deal with illocution constructively

Why do we need the illocutionary facet in order to make sense of linguistic data? In other terms, why morphosyntax and semantics are not enough? Let us explain it through a counterexample, in order to show what kind of information cannot be reduced to morphosyntax and semantics.

(6-1.) you silly

According to the British National Corpus, there are 77 occurrences of *you silly* (6-1). The adpositional tree of this idiomatic expression does not change within them, nonetheless it can be *used* as a vehicle of different information.

(6-1a.) Of course not, you silly.

(6-1b.) Yes you have, you silly.

(6-1c.) Oh, you silly.

All instances of (6-1) are real-case usage of *you silly* reported in the British National Corpus. In particular, (6-1a) is fictional prose, while (6-1bc) are taken from spoken conversations. [7]

What kind of information is different between the three instances? Assumptions, beliefs, desires, willings, attitudes, and any other possible mental state [8] of the speaker on the listener, that the speaker wants to transmit to the listener.

In other words, there is a pragmatic information such as ‘I have mental states such and such on you and I want that you know them’. In (6-1a), the intended meaning perhaps is something like ‘you are likely to presume too much being silly, and I am not happy for that’ while in (6-1b) the speaker’s belief seems to be the opposite: ‘you are likely to forget too much being silly, and I am not happy for that’. Finally, in (6-1c) pragmatic information is probably of a completely different kind: ‘you are likely to forget something being silly, and this fact raises a pleasant feeling in me toward you’ It is obvious that all this information is relevant and it is neither morphological, nor syntactic, nor semantic *sensu stricto*.

Of course, we have said ‘perhaps’, ‘seems’ and ‘probably’ because pragmatic information is vague, i.e., not easily captured by formal means, and exactly this vagueness renders natural languages used by human beings so difficult and, thus, so interesting to analyse. [9]

Nevertheless, a part of pragmatics can be found in adtrees, more precisely in information prominence, as already mentioned in (2-1). In particular, spoken emphasis (put in bold, in (6-2)) can change the trajectory of information prominence (Figure 6.1), but at the same time it gives additional pragmatic information.

(6-2a.) I've seen Mary with Paul. (*no emphasis*)

(6-2b.) I've seen **Mary** with Paul. (*e.g., not Liza*)

(6-2c.) I've seen Mary with **Paul**. (*e.g., not with John*)

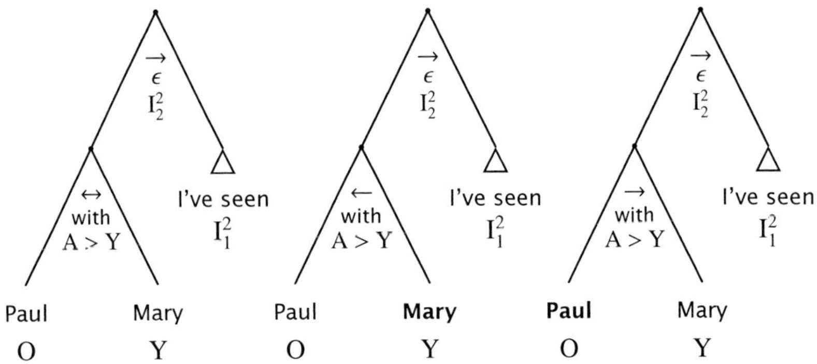

Figure 6.1: How emphasis is mapped into information prominence

In fact, while in (6-2a), which is the unmarked form, there is no particular assumption, belief or other mental state, in (6-2b) it can be assumed that the speaker expected to see someone else (say, Liza) with Paul, but then it reveals that it was Mary to be with him. Analogously, in (6-2c) Mary was expected to be with someone else, (say Liza), not Paul. This assumptions and changes of beliefs, are communicated to the listener by emphasis—in other words, the speaker assumes that the listener receives his illocutions.

However, the trajectory of information prominence is the tip of the iceberg: the illocutions behind the indicator cannot be captured completely by adtrees. We need another level of abstraction in order to deal with pragmatic phenomena.

6.2 Pragmatic adpositional trees

Illocution can be expressed linguistically for analytic purposes but the very wording we choose is arbitrary: in other words, to elicit these information is not neutral, but on the contrary is full of choices of how cognitively and linguistically the scene is perceived. For this reason, illocution should be represented as packed, hidden information, to be elicited only when really needed. We will indicate pragmatic information—hidden under the adtree—with lambdas (λ). [10]

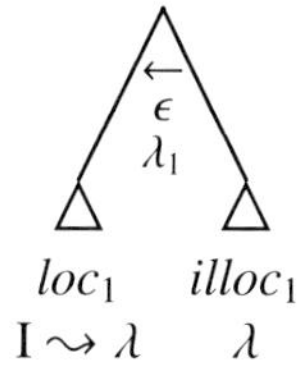

Figure 6.2: Pragmatic abstract adtree at turn 1

Let us imagine that a speaker starts a conversation: we will indicate what is meant in turn 1 with $illoc_1$ and what is actually said with loc_1 (Figure 6.2).

What said (locutionary facet, i.e., loc_1) is governed by what is meant (illocutionary facet, i.e., $illoc_1$): intuitively, humans say something lead by something they mean. Normally, what is intended is more prominent than what is actually said: therefore, the locution-illocution will be government ($\rightarrow$).

If the contrary is true, i.e., if the locutionary facet is more prominent, wording is forced to be interpreted in its basic sense: this is the case of literalism, which is a minority phenomenon in absolute terms. In cases of literalism, the locution-illocution relation will be of dependency ($\leftarrow$). The lambda of the most prominent facet will be promoted as the lambda of the pragmatic adtree—a sort of "adpositional" pragmatic character, even if the "adposition" is always an epsilon, the illocutionary information being linguistically implicit.

By convention, every phrasal construction (i.e., adtree governed by a verbant) can be vehicle of pragmatic information (*lambda*). A *direct* linguistic act occurs when the two lambdas, belonging respectively to locution and illocution, coincide; otherwise the linguistic act will be *indirect*.

Let us suppose that the speaker has requested a linguistic act—if not so, the conversation would be ended, at least by the moment. For instance, in the train the expression *tickets, please!* requires as the default answer a non-linguistic act, i.e., the showing of the appropriate ticket. For the sake of the argument, let us suppose that the speaker requires another speech act at the second turn: [11] does the second turn satisfy the expectations of the first turn speaker? Figure 6.3 shows that a comparison can be done between the second-turn locution and the first-turn illocution ($loc_2? = illoc_1$). The next turns are similar.

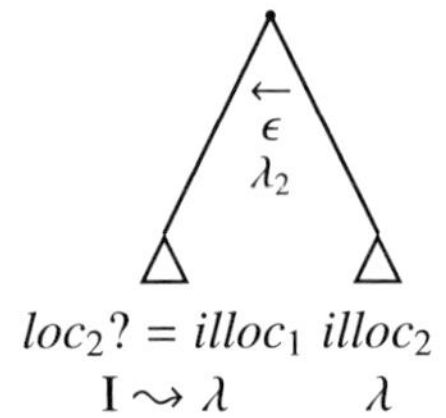

Figure 6.3: Pragmatic abstract adtree at turn 2

Of course, each turn can be formed by more than one locutionary act—i.e., phrasal constructions seen at the pragmatic level—and the correspondent illocutionary facets can change within the same turn. In such cases, the listener can satisfy the last illocutionary facet, or a previous one—or no one. What is important to note is that pragmatics is simply built upon the same rules we have seen in the previous chapters (Figure 6.4).

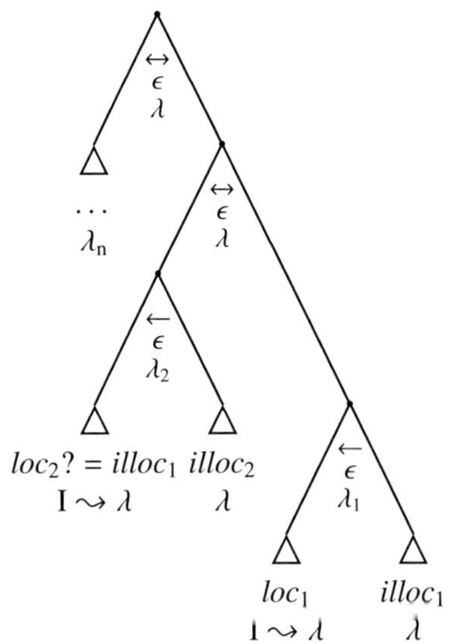

Figure 6.4: How turns are represented

From a formal point of view, nothing changes: lambdas are only further decorations of existing adtrees—hidden by default in case of illocutions, mapped onto verbants in case of locutions. Therefore, what we need now is a taxonomy of illocution, so that we can answer the following question: how many instances of lambdas are there?

6.3 A taxonomy of pragmatic characters

Our aim is to label each phrase (i.e., phrasal construction governed by a verbant) with three lambdas: one for locution, one for illocution, and one of the locution-illocution adtree, by promotion of the most prominent—typically, the illocutionary lambda. It is important to underline that our context is not only theoretical but also practical and applicative: the labelling should be easy to be performed by a trained therapist on the transcript of a conversation, with clear indications on how to proceed without problems of subjectivity. In other words, the taxonomy should be precise and clear beyond individual's opinions on one hand, and practical enough to be used with the appropriate effort.

A relevant part of the philosophical investigation by Searle in the last four decades is devoted to the definition, application and refinement of his taxonomy of illocution and its consequences at philosophical, cognitive, linguistic and social levels. [12] In our perspective, Searle's taxonomy is at a pragmatic level what Tesnière's taxonomy is at a syntactic level: a useful set of characters in order to label locution, illocution and their relations.

Directions of fit

We can describe illocution as the collection of intentional states related to the locution. Searle insists on the fact that "in general, beliefs, desires, and so on are not attitudes to propositions. If I believe that Washington was the first president, my attitude is to Washington and not to the proposition." (Searle, 2010, 27) This is a very important point, because it states that the proposition—in our terms, one or more phrasal constructions somehow linked together—should be analysed to find the expressions of reference (i.e., the statives) and their predications (i.e., verbants). In particular, there are particular verbants whose pragmatic role is their *responsibility* to fit what Austin called the "illocutionary force" to the world.

Let us imagine that the *world*, i.e., the object of a person's perception through senses and mind (pertaining to phenomena or fiction as well) stands below us, while *mind*, i.e., the illocutionary force that drives our mental states and so the propositional locution, stands above us. [13] There are two basic directions between the world and the mind: either downward (mind-to-world) or upward (world-to-mind).

> Thus the aim of a belief is to be true, and it fails if it is false. Insofar as it is true, we can say that the belief matches, or fits, or accurately represents, the world. It has the mind-to-world direction of fit ↓. Desires and intentions, on the other hand, are not supposed to represent how the world is but how we would like it to be (in the case of desires) or how we intend to make it be (in the case of intentions). We can say in such cases that the intention and the desire have the *word-to-mind* direction of fit ↑. (Searle, 2010, 27–28, emphasis in the original).

Of course, beliefs, desires and intentions do not exhaust all the possible mental states. Most emotional states, such as pride, shame, worry, anxiety, gratitude, love, fear, enthusiasm, disgust and so on *presuppose* the existence of the fit.

> Thus if I am proud that I have a big nose or I am ashamed that I have a big nose, in both cases the fact that I have a big nose is simply taken for granted. That is, it is not the aim of the intentional state to represent the fact that I have a big nose (mind-to-world ↓), nor is it its aim to bring it about that I have a big nose (world-to-mind ↑). In these states we simply *presuppose* that I have a big nose (Searle, 2010, 29, emphasis in the original).

Searle (2010) proposes to indicate the presupposed direction of fit with "Presup Fit", or "Presup" for short, even if he admits that such expressions are ugly. In our context, these very long symbols are uncomfortable, as we need to insert them into the pragmatic decoration of adpositional trees. So, we have decided to adopt the symbol = for cases of presupposed direction of fit.

Finally, there are performative-oriented speech acts that *creates* the fit between the mind and the world when they are performed under certain conditions. This kind of speech acts are called Declarations, and they combine the two directions of fit in one: *mind-&-world* (↕).

> The definition is that an utterance is a declaration if the successful performance of the speech act is sufficient to bring about the fit between words [i.e., mind, authors' note] and world, to make the propositional content true. Declarations thus have the double direction of fit ↕ [...] If somebody says "The meeting is adjourned", "I pronounce you husband and wife", "War is declared," or "You're fired," he may succeed in changing the world in the ways specified in these utterances just by performing the relevant speech acts. [...] Fairy stories, by the way, are full of declarations performed by

> witches, wizards, magicians, and so forth. We ordinary humans do not have the ability to perform supernatural declarations, but we do have a quasi-magical power nonetheless of bringing about changes in the world through our utterances; and we are given this power by a kind of human agreement. [...] When the chairman says "The meeting is adjourned," he performs a linguistic *act*, but the *fact* he creates, that the meeting is adjourned, is not a *linguistic fact*. (Searle, 2002, 168–171).

Searle's strong thesis is that Declarations are the main instrument in our construction of social reality, our making of the social world (Searle, 2010, 1995). Table 6.1 subsumes Searle's reflections about the responsibility of the illocutory facet of predication in our linguistic propositions. [14]

The taxonomy proposed by Searle to comprehend all the possible types of speech acts along their illocutionary facet has five elements: assertives ($\downarrow \mathcal{A}$), directives ($\uparrow \mathcal{D}$), commissives ($\uparrow \mathcal{C}$), expressives ($= \mathcal{E}$), Declarations ($\updownarrow \mathbb{D}$).

In constructive adpositional grammars, each element of this taxonomy is a *pragmatic character*, which labels the active verbant of the corresponding phrasal construction (locutionary character, λ_{loc}) and its illocutionary governor (illocutionary character, λ_{illoc}), as already shown in Figure 6.2, which illustrates the pragmatic abstract adtree.

In the sequel, we will explain direct acts before, as they are simpler than indirect ones. In fact, direct speech acts have the same value in both lambdas ($\lambda_{illoc} = \lambda_{loc}$).

Table 6.1: Searle's directions of fit

direction of fit	*symbol*	*intentional states*
mind-to-world	$\downarrow$	beliefs
		visual experience
		remembering
world-to-mind	$\uparrow$	intentions
		desires
mind-world	$=$	pride
presupposed		ashame
		gratitude
		etc.
mind-&-world	$\updownarrow$	*none*

Assertives

Assertives are the first type in Searle's taxonomy, and they give an account of mind-to-world ↓ direction of fit. For this reason, their symbol is the following one: ↓ $\mathcal{A}$, and their prototypical verbant is *assert*. Assertives comprehend statements, descriptions, assertions and so on. They are projected on the ongoing, still in progress or finished (the 'yet').

Each pragmatic character has an exact analogue in intentional states: corresponding to assertives are *beliefs*. Therefore, their essential feature is to be true or false, and their illocutionary content is the proposition itself. In other words, the speaker has evidence (reasons, etc.) for the truth of the proposition, and he communicates this belief to the listener as well. There is no attempt to convince the listener to adopt the speaker's belief. "Thus 'I am simply stating that *p* and not attempting to convince you' is acceptable" (Searle, 1969, 66).

(6-3a.) I won the race (*I am telling that I won the race*)

Let suppose that a champion in Formula 1 racing is giving an interview (6-3a)[15]. There are two *pragmatic* actants: the first actant is the speaker S, in this case the champion, while the second actant is the speaker's belief (Y, S), where here Y stands for the belief.

(6-4.) *Speaker asserts that speaker's belief exists*

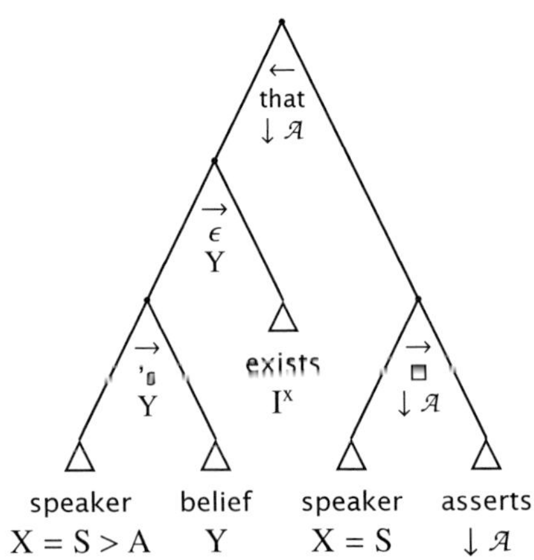

Figure 6.5: Abstract pragmatic adtree for assertives

Example (6-4) is a possible wording for the abstract pragmatic adtree for assertives (Figure 6.5). The illocution stands at the right of the adtree (*speaker asserts*), and the adposition was boxed as it does not make sense to get it explicit.

On the left side there is the locutionary part, where the *morphosyntactic* valency was left underspecified (I^x), as the two *pragmatic* actants of assertives are completely independent from the actual, morphosyntactic realisations performed by locution. This important fact is easy to show through an example.

(6-5.) I tell you what I found in the library.

For instance, (6-5) is clearly an assertion, and its locution is governed by the trivalent verbant *tell*: *I* is the first actant, *what I found in the library* taken as a whole, is the second actant, while finally *you* is the third one.

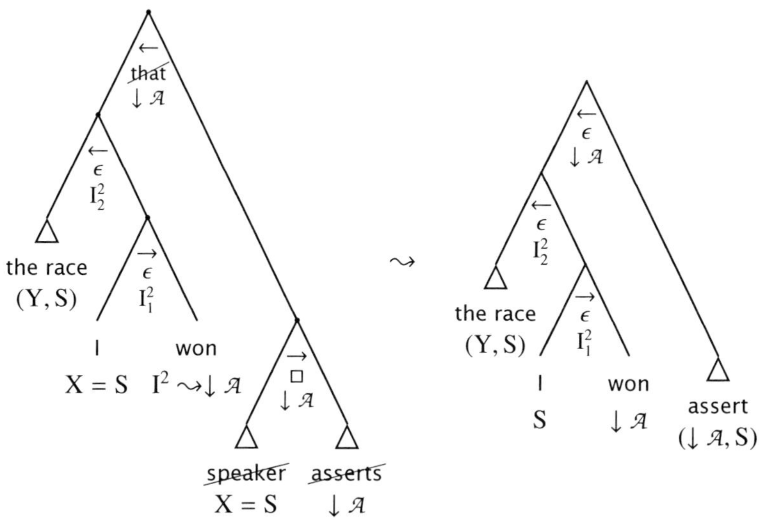

Figure 6.6: Pragmatic adtree of the assertive *I won the race*

Figure 6.6 shows how to instantiate the abstract pragmatic adtree for the assertive (6-3a). While on the left adtree some morphosyntactic information was provided in order to help the reader in establishing the bridge between constructive morphosyntax and pragmatics, on the right only the pragmatic labels were made explicit: in particular, (↓ 𝒜, S) indicates in a compact way that illocution is performed by S.

Directives

The aim of directives is to try to get other people to do things. The speaker who says the utterance is not satisfied of the world as it is, so it tries to change it. Therefore, directive direction of fit is world-to-mind ↑.

Prototypical examples of directives are: orders, commands, requests, but also desires, wants, wishes and so on. Their symbol is $\uparrow \mathcal{D}$, and their prototypical verbant is direct. The intentional state which corresponds to directives is *desire*, thus it has no sense to ask if a directive is true or false, because the illocutionary content is the performative request made to the receiver.

Therefore, unlike assertives, directives have three pragmatic actants: the speaker S who acts as the requester, the receiver R and the actual request to be performed by the receiver (Z, R).

(6-6.) *Speaker directs receiver to do request*

Example (6-6) shows a possible abstract wording for directives (Figure 6.7). Imperatives are explicit directives, such as in (6-3b), where the team leader asks the champion to win the Formula 1 race (Figure 6.8).

(6-3b.) Win the race! (*I order you to win the race*)

Sometimes the illocutive verbant is expressed explicitly—this is valid for any pragmatic character, not only directives, as in (6-7).

(6-7.) Liza ordered Paul to leave.

In such cases, the illocutionary part of the adtree can be rendered explicitly as in Figure 6.9.

Commissives

Directives and commissives share the property of having an upward direction of fit ↑, and for this reasons they are called collectively *motivators* (Searle, 2010, 126). Unlike assertives, they are both projected to the future (the 'not yet').

However, while directives are trying to change the world through others, commissives rely on the speaker himself. Typical commissives are promises, vows and pledges. Their intentional state is pure, i.e., it is *intention*. The conventional symbol for commissive is $\uparrow C$, and the prototypical verbant is *commit*.

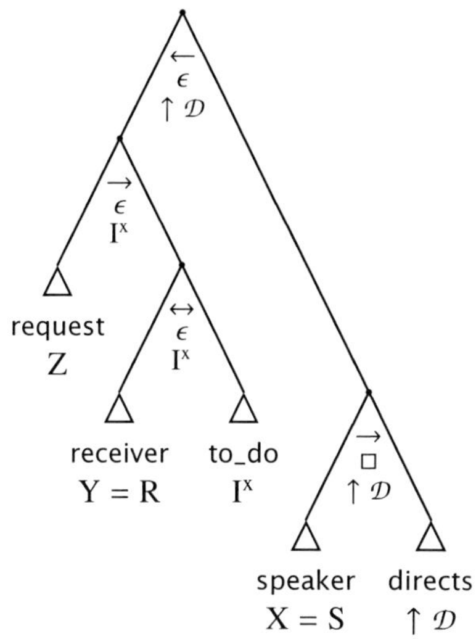

Figure 6.7: Abstract pragmatic adtree for directives

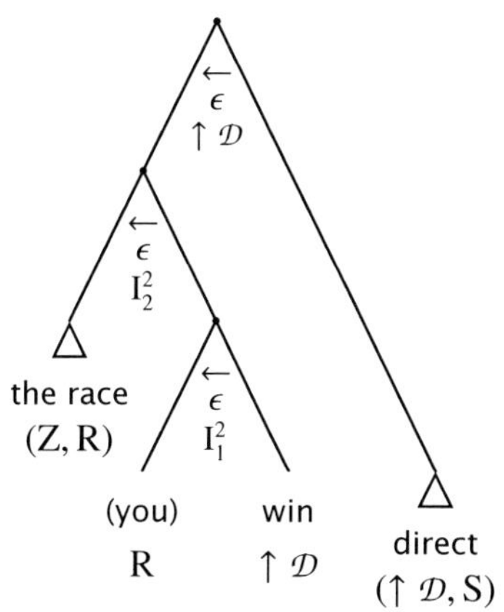

Figure 6.8: Pragmatic adtree of the directive *Win the race!*

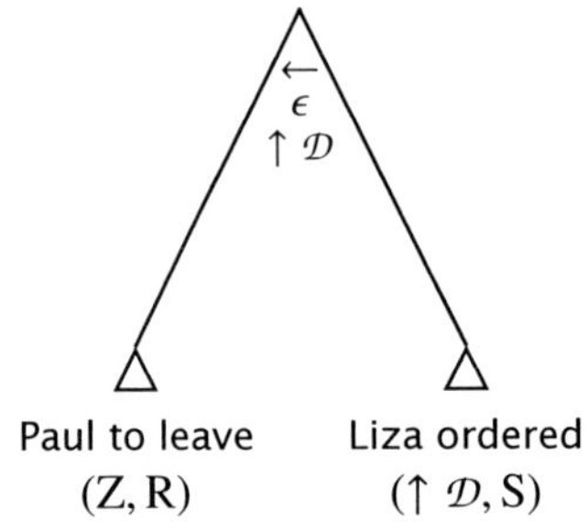

Figure 6.9: Pragmatic adtree of the directive *Liza ordered Paul to leave*

Example (6-8) is a possible wording of commissives (Figure 6.10): the analogies and differences with directives are quite clear.

(6-8.) *Speaker promises that speaker will do the promise*

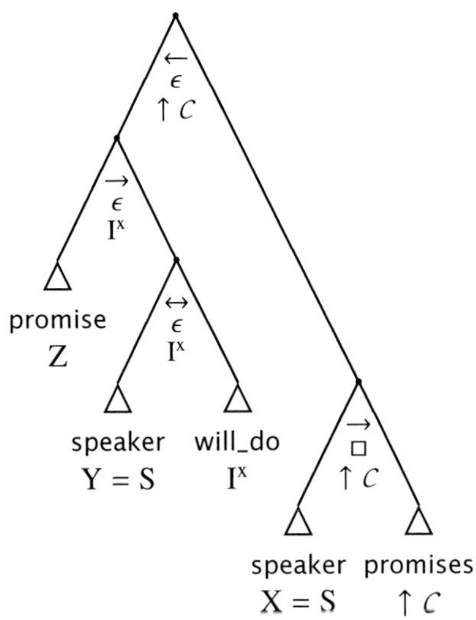

Figure 6.10: Abstract pragmatic adtree for commissives

If we expand the scene around the Formula 1 racing stated before, in (6-3c) is reported what the champion says, in his car, before the start (Figure 6.11).

(6-3c.) I'll win the race (*I promise myself to win the race*)

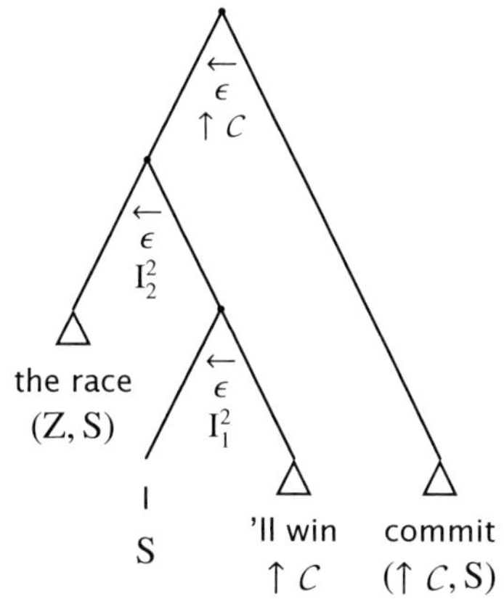

Figure 6.11: Pragmatic adtree of the commissive *I'll win the race*

It is important to note that a commissive is always referred to the speaker S: properly, the pure intention is not directed to others. In (6-3c), the champion is *not necessarily* entangling himself with an answer to his team leader's directive—even if such a scenario is not only possible but also realistic. In such a case, the intention is no more pure, even if the pragmatic character remains $\uparrow \mathcal{C}$. In fact, most utterances do not live in isolation, but in conversations, where a complex architecture of implications, presuppositions, assumptions, are continually negotiated among the actors of the conversation itself.

Expressives

Speakers use expressives when the relation mind-world is taken as granted, is presupposed, and some feeling, attitude or emotion about a state of affairs is needed to be expressed. The conventional symbol of expressive is $= \mathcal{E}$, and it covers the whole range of intentional states where the emotional part is relevant. Searle's examples are greetings, congratulations and apologies, but the list can be extended easily, following cultural contexts, time, age, personality of speakers, and so on. The prototypical verbant is *express*.

(6-9.) *Speaker expresses speaker's emotion on receiver*

Figure 6.12 shows the pragmatic adtree of a possible abstract wording of expressives (6-9). Like commissives, the intentional state of expressives stands in the speaker, but unlike commissives—and like directives—the third actant is the receiver R of the speaker's emotion (Y, S). Sometimes (Y, S) is not entirely known even by the speaker himself: this is part of the reason while human beings are introspective, and hence why psychologists and therapists in general do exist in *any* human society—call them shamans, rabbis, priests, or whatever. [16]

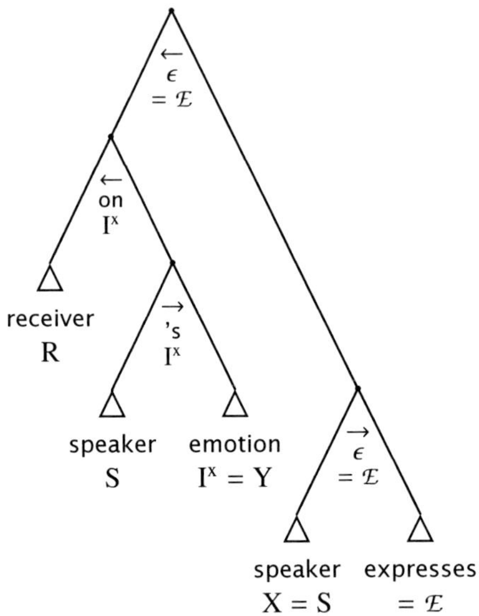

Figure 6.12: Abstract pragmatic adtree for expressives

Gratitudes are typical expressives. A thank is always for an act done by someone different from the speaker after something happened. There is neither intention to modify the world nor to assert something about the world: gratitude takes what happened as granted.

(6-3d.) I thank my team for their help.

Example (6-3d) is an utterance said during an interview by the champion just after the end of his race, which he had won, and before the official proclamation. Its pragmatic character is expressive, and its adtree is illustrated in Figure 6.13.

Expressives can be considered the residual character of the whole taxonomy, as it comprehends all "other" mental states, different from beliefs, desires and intentions—Declarations having no intentional states at all. Nonetheless, their

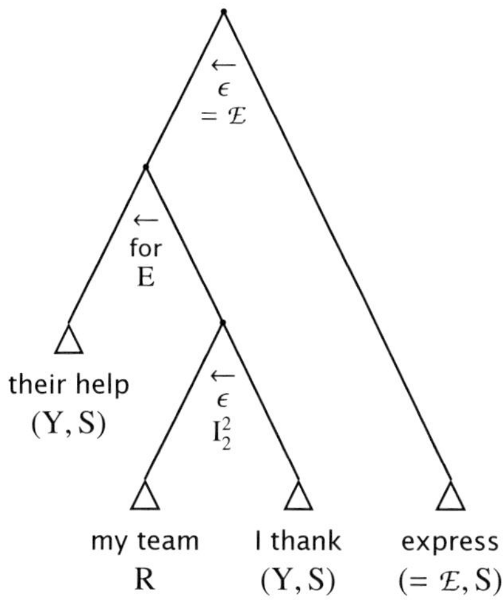

Figure 6.13: Pragmatic adtree of the expressive *I thank my team for their help*

importance cannot be put below assertives, directives or commissives. Moreover, expressives are very important for therapists, as they permit to colour the pragmatic picture with a full palette of emotions.

Declarations

Searle (2010) devotes most of his reflections on Declarations, because they create, establish and transform social reality. Their direction of fit is both mind-to-world and world-to-mind, i.e., mind-&-world ↕ and hence their symbol is the following: ↕ $\mathbb{D}$.[17] What Declared is a reality built *over* the physical one.

> Our sense that there is an element of magic, a conjuring trick, a sleight of hand in the creation of institutional facts out of brute facts derives from the nonphysical, non-causal character of the relation of the X and Y terms in the structure where we simply *count* X things as Y things. In out toughest metaphysical mood we want to ask 'But is an X really a Y?' For example, are these bits of paper really *money*? Is this piece of land really somebody's *private property*? Is making noises through the mount really *getting married*? [...] We do not have this sense of giddiness where the agentive function is performed entirely in virtue of physical features. This, we do not have any metaphysical doubts about whether or not this is really

> a screwdriver, or this is really a car, because the sheer physical features of the objects in question enable them to function as screwdrivers or cars. (Searle, 1995, 45)

Declarations are special in many ways. First, there is no intentional state because the Declaration is performative: instead of posing an intent, the speaker performs an act.

Declarations live in the 'here and now' aspect of time: before the Declaration, some social reality did not exist, after the Declaration, it does exist.

(6-3e.) ...and the winner of the race is...

In the example (6-3e) the champion is proclaimed the winner of the race through a Declaration.

There are strict requisites to have a Declaration, where most of them are truly extra-linguistic. Searle calls them collectively the *activating context* (our symbol: $\mathbb{C}$) which is analysed as such: (a) an extra-linguistic institution; (b) a special position by the speaker within the institution; (c) another position by the receiver, again within the institution; (d) a special convention that certain literal sentences of the given natural language count as performances of giving the social status $\mathbb{S}$ and (e) the intention by the speaker in the utterance to have a Declarative status (Searle, 2002, see at least 169).

For instance, in (6-3e) the extra-linguistic institution is the Formula 1 World Drivers' Championship (WDC, recognised by a set of Declarations), while the speaker S is the official referee in this race – status given by another Declaration. The receiver R should be the driver who actually won the race, and through the speaking of (6-3e) he acquires the new social status $\mathbb{S}$ becoming the winner of the race. All components of $\mathbb{C}$ are needed in order to activate the Declaration: for instance, if the speaker is not the official referee but a champion's fan, the Declaration fails $\not{\updownarrow}\not{\mathbb{D}}$.

Declarations live within institutions: for instance, within his family he is the father (for his children) and the husband (for his wife), while for his old good friends of his college he is simply an alumnus.

Institutions are social realities which collectively form a society; within their society, institutions held *deontic powers*: someone has some *right* to do things in holding the status $\mathbb{S}$ and some *obligations* as well. For example, religion as such is not an institution, while the Catholic Church is.

There is a test to understand if a activating context pertains to an institution: "If people do not now believe it, and did not in the past believe it, would it still be true? In the case of a recession, if people didn't believe that it was a recession, it would still be a recession" (Searle, 2010, 117). Sometimes institutions fall, and with them all the Declarations belonging to them. This is often a dramatic fact, as for example the fall of USSR.

A Declaration cannot be retreated, but it can be corrected or changed by another Declaration: if Paul gets married he is no longer a bachelor but he is Declared a husband; he can divorce, but he cannot retreat the Declaration of marriage in order to come back to bachelorhood. For this reason, Searle (2010) asserts that society is *constructed* from an architecture of Declarations. The ontogeny of such capacity of creation, transformation and maintenance of institutional reality is in children's fictional play, such as "Okay, I'll be Adam, you be Eve, and we'll let this block be the apple," a crucial capacity for sane growing, as pointed out by Rakoczy and Tomasello (Searle, 2010, 121).

(6-10.) *Speaker Declares: 'the Receiver has the Declared Status'*

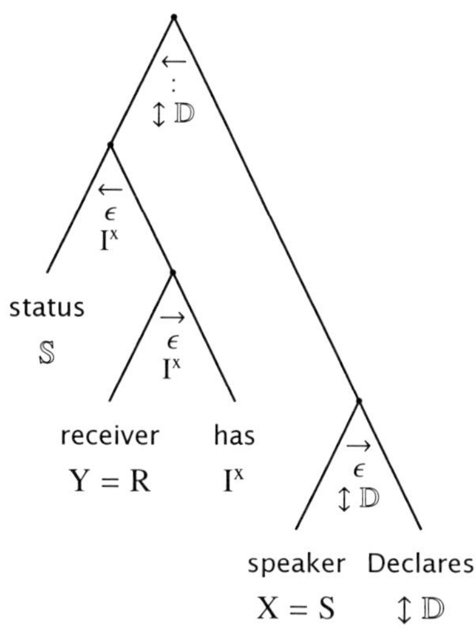

Figure 6.14: Abstract pragmatic adtree for Declarations

Example (6-10) is a possible wording for Declarations. Institutions are not only governmental or economic: local sporting clubs, schools, private property, friendship, family are all valid institutions, as the test for the activating context is passed—i.e., the object named has deontic powers.

(6-11.) (*Father:*) You (*son*) are the black sheep of the family.

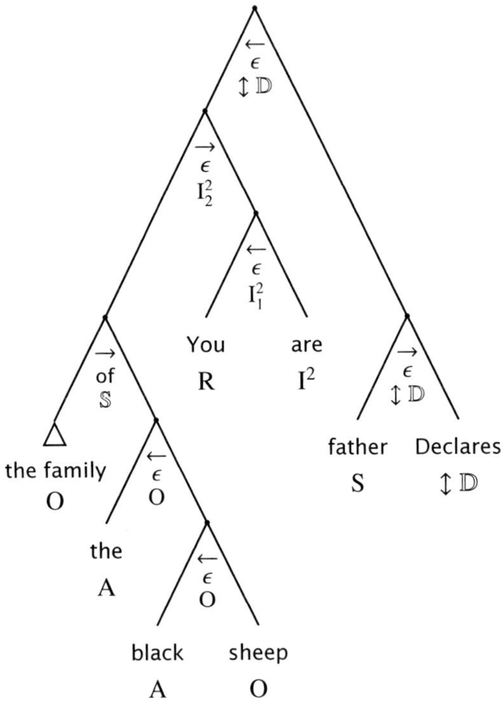

Figure 6.15: Pragmatic adtree of ...*black sheep*...

In (6-11) the speaker S is a father who Declares to the receiver R, his son, that he owns the status $\mathbb{S}$ of being *the black sheep of the family*—counting as a whole under a pragmatic point of view (Figure 6.15). In (6-11) the institution within the activating context $\mathbb{C}$ is valid and it is *the family*: unlike other examples, in (6-11) the relevant institution is said explicitly. Of course, a father can Declare it successfully because he owns a proper status in the family. If the speaker were a son's friend, for instance, (6-11) would be simply an assertive ($\downarrow \mathcal{A}$).

Our claim is that Constructive Conversational Analysis can help a therapist in making explicit the institutions and related Declarations that rules patient's life, quite often him being unaware of them.

Table 6.2: The pragmatic characters

type of illocution	*verbant*	*symbol*	*intentional state*	*direction of fit*	*pragmatic actants*
assertive	*assert*	$\downarrow \mathcal{A}$	beliefs	mind-to-world	S; (Y, S)
directives	*direct*	$\uparrow \mathcal{D}$	desires	world-to-mind	S; R; (Z, R)
commissives	*commit*	$\uparrow \mathcal{C}$	intention	world-to-mind	S; S; (Z, S)
expressives	*express*	$= \mathcal{E}$	*many*	Presup	S; (Y, S); R
Declarations	*Declare*	$\updownarrow \mathbb{D}$	*none*	mind-&-world	S; ($\mathbb{S}$, R)

6.4 Constructive indirect speech acts

Table 6.2 shows pragmatic characters in comparison. Each pragmatic character is structurally different, either of its direction of fit, or the pragmatic actants, and so on. In particular, Declarations require the activating context $\mathbb{C}$ which is out of the linguistic act. Furthermore, while it is possible to form morphosyntactic questions which are pragmatically assertives (6-12), directives (6-13), commissives (6-14) and expressives (6-15), it is not possible to Declare a question or to question a Declaration, for obvious reasons: the making of social reality is a quasi-magic act, which literally cannot be questionable.

(6-12.) Did you see the news? ($\downarrow \mathcal{A}$)

(6-13.) Why have you still to finish your homework? ($\uparrow \mathcal{D}$)

(6-14.) On what I have my mind fixed today? ($\uparrow \mathcal{C}$)

(6-15.) How should I feel about that? ($= \mathcal{E}$)

Until now, we have seen direct acts, i.e., where the pragmatic character of the illocution is the same of the pragmatic character of the locution ($\lambda_{illoc} = \lambda_{loc}$). In truth, real-world speech acts are often indirect. For instance, typical indirect acts are commissives masked as directives and vice versa.

(6-16.) I promise you'll do it.

(6-17.) I request myself to be brave.

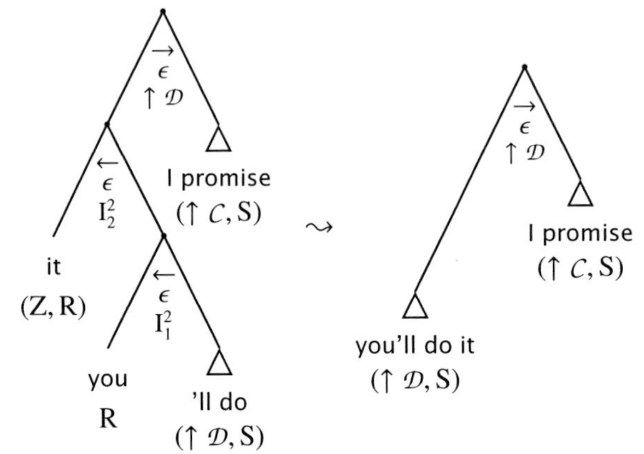

Figure 6.16: Adtrees of the indirect directive *I promise you'll do it*

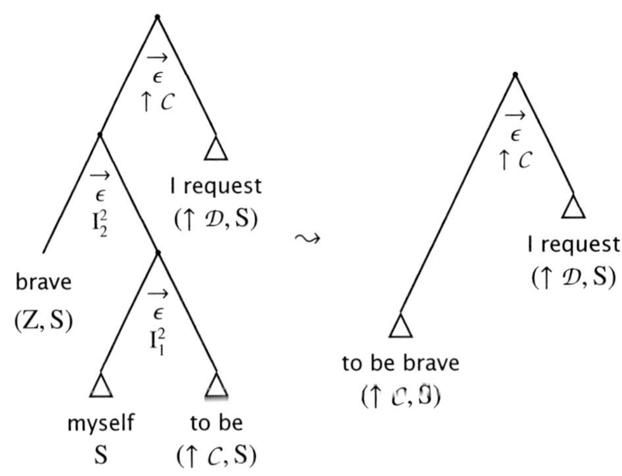

Figure 6.17: Adtrees of the indirect commissive *I request myself to be brave*

The governor *I promise* of (6-16) is prototypical of commissives, but the inner structure of the dependent (*you'll do it*) reveal the stronger pragmatic character of directive (Figure 6.16). In fact, (6-16) seems likely to be a sort of menace: the final pragmatic character will be directive.

(6-18.) Why should I be angry with him? (= $\mathcal{E} \rightsquigarrow\uparrow \mathcal{D}$)

(6-19.) I wonder if I am ashamed to follow her. (= $\mathcal{E} \rightsquigarrow\downarrow \mathcal{A}$)

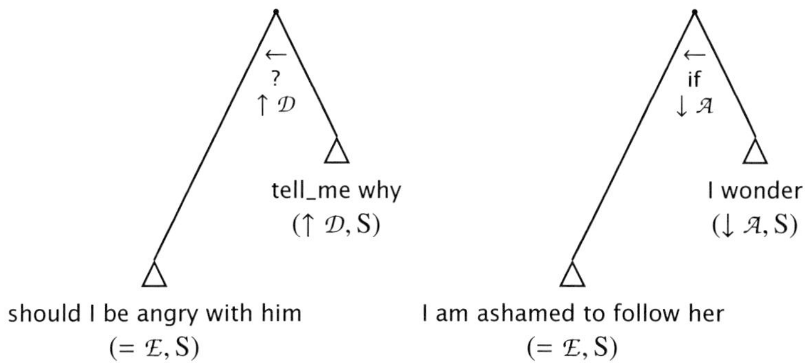

Figure 6.18: Two examples of indirect acts encapsulating expressives

In principle, any pragmatic character can be under disguise. Let us present two examples of expressives. (6-18) is a request $\uparrow \mathcal{D}$ to the listener about a feeling of anger, which is an expressive $= \mathcal{E}$. Moreover, (6-19) encapsulates the ashame $= \mathcal{E}$ into a belief through the governor *I wonder*, so that the final pragmatic character is assertive $\downarrow \mathcal{A}$ (Figure 6.18). Again, unlike the others types, Declarations seem not to be suitable for masquerade or indirectness.

6.5 Constructive Conversational Analysis

Until now, Constructive Conversation Analysis (CoCAL) has been proposed for the analysis of transcripts in therapeutic settings. A consolidation of CoCAL through experimental evidence is out of the scope of the present book; however, it is still possible to perform an analysis on a couple of real-world examples to give a cue of the direction we want to take in this area. The example of application is the analysis an extract from one of the interactions between Stephen E. Finn, the founder of Therapeutic Assessment, and Jim, one of his clients. [18]

A client-therapist interaction fragment

The fragment of the interaction is 8 turns long, which means less than a minute of conversation (Finn, 2007, 93). Let us consider that a client-therapist interaction normally lasts approximately 45 minutes; therefore, our data are very limited. Nonetheless, some useful information can still be extracted.

t01:	*Steve*:	I wonder if you're willing to do a little experiment with me?
t02:	*Jim*:	What's that?
t03:	*Steve*:	I'd like to give you another achievement task that would normally pull for the loser lens, and see if we can keep it from happening.
t04:	*Jim*:	OK. How would we do that?
t05:	*Steve*:	Do you have any ideas based on what we talked about?
t06:	*Jim*:	Well, I guess first I should just keep going, no matter how badly I think I'm going.
t07:	*Steve*:	Good idea. And what might that be like?
t08:	*Jim*:	I guess I'll feel anxious.

At first, the turns can be rewritten in order to extract the information relevant for the therapy: the illocutionary and locutionary pragmatic characters of each phrasal construction. In this way, direct and indirect speech acts can be analysed in their components for further observations, in particular to individuate conversational patterns between the client and the therapist.

Table 6.3 is a possible extraction of pragmatic information from the extract (Finn, 2007, 93). The possible *pragmatic* actants are listed for needs of completion, even if sometimes they cannot be applied (□). The most important column is the rightmost one, where the final, illocutionary pragmatic character λ_{illoc} is provided. Of course, it is always possible to get the pragmatic adtrees—built upon the morphosyntactic adtrees as well—explicit under request, but the feedback received by the first applications of CoCAL we had with psychotherapists is that the transcript, a table and a comment written in natural language is adequate for a good report.

Mainly, the fragment is based on assertives–with Steve explaining the *little experiment*, i.e. the task proposed to Jim—and directives—will Jim accept to do the task, as requested by Steve? In turn 1 the therapist makes a request $\uparrow \mathcal{D}$ under a belief $\downarrow \mathcal{A}$, but the answer in turn 2 is not a commissive, but rather a request for clarification $\uparrow \mathcal{D}$. So, in turn 3a the therapist softens the request $\uparrow \mathcal{D}$, explaining

Table 6.3: The Constructive Conversational Analysis of Finn (2007, 93)

turn	*speaker*	*verbant*	*1st act.*	*2st act.*	*3st act.*	λ_{loc}	λ_{illoc}
t01	Steve	wonder	Steve	task	□	$\downarrow \mathcal{A}$	$\uparrow \mathcal{D}$
t02	Jim	is	task	□	□	$\uparrow \mathcal{D}$	$\uparrow \mathcal{D}$
t03a	Steve	like give	Steve	task	Jim	$\downarrow \mathcal{A}$	$\uparrow \mathcal{D}$
t03b	Steve	would pull	task	□	□	$\downarrow \mathcal{A}$	$\downarrow \mathcal{A}$
t03c	Steve	see if	we	task	□	$\downarrow \mathcal{A}$	$\downarrow \mathcal{A}$
t04a	Jim	OK	Jim	task	□	$\downarrow \mathcal{A}$	$\uparrow \mathcal{C}$
t04b	Jim	would do	we	task	□	$\uparrow \mathcal{D}$	$\uparrow \mathcal{D}$
t05	Steve	have	Jim	ideas	□	$\downarrow \mathcal{A}$	$\uparrow \mathcal{D}$
t06	Jim	guess	Jim	task	□	$\uparrow \mathcal{C}$	$\downarrow \mathcal{A}$
t07a	Steve	good idea	Steve	□	□	$= \mathcal{E}$	$\downarrow \mathcal{A}$
t07b	Steve	might like	task	□	□	$\downarrow \mathcal{A}$	$\uparrow \mathcal{D}$
t08a	Jim	guess	Jim	□	□	$\downarrow \mathcal{A}$	$\downarrow \mathcal{A}$
t08b	Jim	feel	Jim	anxious	□	$= \mathcal{E}$	$= \mathcal{E}$

why he believes ($\downarrow \mathcal{A}$, turn 3bc) that the request $\uparrow \mathcal{D}$ is useful and appropriate for the client.

The turning point is in turn 4a, when Jim says *OK*, accepting to take in charge the request as a commitment $\uparrow \mathcal{C}$: in fact, he uses the pronoun *we*, i.e., Jim and Steve together. In other words, the task is performed by Jim, but Jim is not alone in doing it. The request for clarification $\uparrow \mathcal{D}$ in turn 4b is obvious.

Turn 5 shows a different kind of request $\uparrow \mathcal{D}$: Steve wants to know Jim's beliefs $\downarrow \mathcal{A}$ before to start with the task itself, and Jim answers in the following turn 6. In turn 7 the therapist approves, confirms, both at a level of beliefs and of emotions $= \mathcal{E}$, what expected by Jim, but he is still unsatisfied as he requests $\uparrow \mathcal{D}$ Jim's feelings about the task, and in fact Jim tells he *feels anxious* $= \mathcal{E}$ in turn 8. It is quite clear that the conversational strategy by the therapist is to get Jim comfortable, on one hand, and to make him say his anxiety, on the other hand, in a delicate equilibrium.

CHAPTER SEVEN

CONCLUSION

The adpositional paradigm in constructive linguistics is still in its infancy stage, but nevertheless its expressive power is promising. In our view, there are two possible further directions in developing the paradigm in general and the model presented here in particular: first, delving in the mathematical implications of the formal model could give unexpected insights on natural language grammar considered as a space, or a logic; second, the description of *any* linguistic phenomena in terms of morphemes, constructions and transformations. Our aim is to develop these two further directions in parallel, as most results presented in this book were obtained by intertwining the linguistic and mathematical perspectives.

Wherever ordered written linguistic data are available—e.g., corpora obtained from transcripts of therapeutic settings, television or radio dialogues, or even literary criticism, prose, poetry and theatre—Constructive Adpositional Grammar analysis can be performed. Literary critics, scholars belonging to humanities in general—as well as philosophers of language—can apply the model to show the results of their efforts in a new and unusual way which has the advantage of being formally sound.

In fact, Constructive Conversational Analysis (CoCAL) showed that the adpositional paradigm and in particular the linguistic and formal model presented in this book can be tuned for specific needs and concrete applications: some details can be left hidden—because not important for the specific purpose—but at the same time they can be recovered under request.

By the moment, experimental evidence of CoCAL is still lacking, but we are confident that results will be available when the fruitful collaboration with a research team in psychotherapy will be continuing.

In general, whenever a linguistic corpus of written text is produced, in principle Constructive Adpositional Grammars can be used. The morphosyntactic basis gives also an account of semantics, while pragmatics can be rendered through Searle's pragmatic characters or other perspectives. [1]

A further application proposal is constructive criticism, when a fixed set of textual roles to actants is given to analyse texts. For instance, Greimas' Actantial Model could be another possible decoration of adpositional trees for the purposes of literary analysis and criticism (Greimas, 1966).

The adpositional paradigm is flexible and precise at the same time. Our tenet is that natural language grammars can be represented in a formal, mathematical way, such that their *meaning* can be generated within the model itself.

Meaning is more than semantics: in fact, semantics is born from the relations among sememes, and meaning activates semantics instantiating sememes and their relations in a given referential world, which can be either real or fictional. In fact, not only we cannot have scenes without reference to a world, but also we cannot have scenes without some sort of representation.

Linguistically, scenes start to exist when we perform a speech act or write our wordings down, and while our listener or reader activates the infinite and indefinite process of sense—what Frege called *Sinn* (sense) and *Bedeutung* (reference). Each pragmatic act produces a meaning cloud that generates—and regenerates—that cultural convention usually called 'natural language'. We track a trace of that process through lambdas: it is still an imperfect way to do it, but nonetheless it is correct. In fact, what is observable in a language is its productions, i.e., the collections of phrasal constructions, their links, i.e., transformations, and their inner constructions.

Our claim is that our description of natural language, which ultimately generates a cloud of meaning and structure over any linguistic expression, can be fruitfully represented inside topos theory, in particular via Grothendieck's topoi. We studied the category of adtrees throughout the whole book. What if the adtrees are considered as denoting objects of a category which is a special model of the world/scene we are speaking of? If we consider that all the governors of a set adtrees form a sieve, and all the admitted constructions starting from a given adtree form the cosieve, we see that, mathematically, the world we are speaking of becomes a (generalised) topological space—a *site*—whose spatial structure is describe by the sieves, or, dually, by the cosieves, that is, by the inner structure of the language we are using to describe the world itself. This means that the meaning and the structure of the language projects itself onto the world giving it the shape of a 'space' whose structure is generated, ultimately, by the meaning of the expressions we use to speak of the world.

The linguistic counterpart of this formalisation is almost straightforward: the basic schemata to construe a language are the verbants and their meaning, i.e., the transformations generated by the morphemic instantiation of the adpositional trees themselves. In other words, meaning is *generated* within the model and then mapped onto the scene. We want to underline the fact that this generation is mathematical indeed, but—unlike the classic Chomskyan perspective—this generation cannot be reduced to a computational or algorithmic model. The reverse is true: a computational, algorithmic model can be extracted from the mathematical generation within our topos-theoretical model but what is linguistically represented is bound to be a partial snapshot of a natural language in a given model. In other words, we can generate grammar and meaning but we cannot get this generation completely explicit—not all meaning can be written down, as the topos we deal with is really a big entity, usually a proper class and not merely a set.

Constructive linguistics give a canonical and systemic methodology to put two natural languages in comparison: a constructive linguist would take the two classes of basic sememes—most probably the verbants—and the patterns of transformations of constructions starting from them. In fact, topoi, and in particular Grothendieck's topoi, can be seen as a coordinated collection of information. For this reason, we are deeply convinced that Grothendieck's topoi are the right mathematical instrument to formalise languages and their regularities. After all, symmetry has always a deep reason. If not, this book could not have been written.

APPENDIX A

EXAMPLES

All examples of adpositional trees given throughout the book are presented here as linearised adtrees. Epsilon-reductions are made where appropriate.

A.1 Introduction

(1-1. — page 8) La vecchia porta la sbarra.

$\epsilon^{\rightarrow}_{I^2_2}((\text{la})_{O_2}, \epsilon^{\leftarrow}_{I^2_1}((\text{La vecchia porta})_{O_1}, (\text{sbarra})_{I^2}))$

$\epsilon^{\leftarrow}_{I^2_2}((\text{la sbarra})_{O_2}, \epsilon^{\rightarrow}_{I^2_1}((\text{La vecchia})_{O_1}, (\text{porta})_{I^2}))$

(1-2. — page 9) Time flies like an arrow and fruit flies like a banana.

$\text{and}^{\leftrightarrow}_{I^0}(\text{like}^{\rightarrow}_{I^0}(\epsilon^{\rightarrow}_{O}((\text{an})_A, (\text{arrow})_O), \epsilon^{\leftarrow}_{I^1_1}((\text{time})_O, (\text{flies})_{I^1})),$
$\epsilon^{\leftarrow}_{I^2_2}(\epsilon^{\rightarrow}_{O}((\text{a})_A, (\text{banana})_O), \epsilon^{\leftarrow}_{I^2_1}(\epsilon^{\rightarrow}((\text{fruit})_A, (\text{flies})_O), (\text{like})_{I^2})))$

(1-3. — page 9) È la rossa di Maranello.

$\epsilon^{\leftarrow}_{I^2_2}(\text{di}^{\leftarrow}_{O}(\text{Maranello}_O, \epsilon^{\rightarrow}_{O}(\text{la}_A, \epsilon^{\rightarrow}_{O}((\text{rossa})_A, \square_{O_1}))), \epsilon^{\rightarrow}_{I^2_1}(\square_{O_1}, \text{È}_{I^2}))$

A.2 Syntax

(2-1a. — page 20) I've not seen Liza. She was not with Paul.

$\epsilon^{\leftrightarrow}_{I}((\text{She was not...})_I, (\text{I've not...})_I)$

(2-1b. — page 20) I've **not** seen Liza, she was not with Paul.

$\epsilon^{\leftarrow}_{I}((\text{She was not...})_I, (\text{I've } \textbf{not}\text{...})_I)$

(2-1c. — page 21) I've not seen Liza, she was **not** with Paul.

$\overrightarrow{\epsilon_{I}}$((She was **not**...)$_{I}$, (I've not...)$_{I}$)

(2-2a. — page 24) The bicycle belongs to Kim.

$\overrightarrow{to_{I^2_2}}$(Kim$_{O_2}$, $\overleftarrow{\epsilon_{I^2_1}}$((The bicycle)$_{O_1}$, (belongs)$_{I^2}$))

(2-2b. — page 24) Kim owns the bicycle.

$\overleftarrow{\epsilon_{I^2_2}}$((the bicycle)$_{O_2}$, $\overrightarrow{\epsilon_{I^2_1}}$(Kim$_{O_1}$, (owns)$_{I^2}$))

(2-3a. — page 25) The janitor will open the door with the key.

$\overrightarrow{with_{I^3_3}}$((the key)$_{Z}$, $\overleftarrow{\epsilon_{I^3_2}}$((the door)$_{Y}$, $\overrightarrow{\epsilon_{I^3_1}}$((the janitor)$_{X}$, (will open)$_{I^3}$)))

(2-3b. — page 26) The janitor will open the door.

$\overleftarrow{\epsilon_{I^2_2}}$((the door)$_{Y}$, $\overrightarrow{\epsilon_{I^2_1}}$((the janitor)$_{X}$, (will open)$_{I^2}$))

(2-3c. — page 26) The key will open the door.

$\overleftarrow{\epsilon_{I^2_2}}$((the door)$_{Y}$, $\overrightarrow{\epsilon_{I^2_1}}$((the key)$_{X}$, (will open)$_{I^2}$))

(2-3d. — page 26) The door will open with the key.

$\overrightarrow{with_{I^3_3}}$((the key)$_{Y}$, $\overleftarrow{\epsilon_{I^3_1}}$((the door)$_{X}$, (will open)$_{I^3}$))

(2-3e. — page 26) The door will be opened with the key.

$\overrightarrow{with_{I^3_3}}$((the key)$_{Y}$, $\overleftarrow{\epsilon_{I^3_1}}$((the door)$_{X}$, (will be opened)$_{I^3}$))

(2-3f. — page 26) The door will be opened by the janitor.

$\overrightarrow{by_{I^2_2}}$((the janitor)$_{Y}$, $\overleftarrow{\epsilon_{I^3_1}}$((the door)$_{X}$, (will be opened)$_{I^2}$))

(2-3g. — page 26) The janitor will open the door with this key for the Lady.

$\overrightarrow{for_{I^0}}$((the lady)$_{O}$, $\overrightarrow{with_{I^3_3}}$((this key)$_{Z}$,
$\overleftarrow{\epsilon_{I^3_2}}$((the door)$_{Y}$, $\overrightarrow{\epsilon_{I^3_1}}$((the janitor)$_{X}$, (will open)$_{I^3}$))))

(2-4a. — page 31) Ann gave Kim the car. (not the bicycle)

$$\epsilon^{\rightarrow}_{I^3_3}((\text{the car})_Z, \epsilon^{\leftarrow}_{I^3_2}(\text{Kim}_Y, \epsilon^{\rightarrow}_{I^3_1}(\text{Ann}_X, \text{gave}_{I^3})))$$

(2-4b. — page 31) Ann gave the car to Kim. (not to Bruce)

$$\text{to}^{\rightarrow}_{I^3_3}(\text{Kim}_Z, \epsilon^{\leftarrow}_{I^3_2}((\text{the car})_Y, \epsilon^{\rightarrow}_{I^3_1}(\text{Ann}_X, \text{gave}_{I^3})))$$

(2-4c. — page 32) Ann gave him the car.

$$\epsilon^{\rightarrow}_{I^3_3}((\text{the car})_Z, \epsilon^{\leftarrow}_{I^3_2}(\text{him}_Y, \epsilon^{\rightarrow}_{I^3_1}(\text{Ann}_X, \text{gave}_{I^3})))$$

(2-5a. — page 32) Maria deu o livro para/a/ϵ Pedro

$$\text{a/para/}\epsilon^{\rightarrow}_{I^3_3}(\text{Pedro}_Z, \epsilon^{\leftarrow}_{I^3_2}((\text{o livro})_Y, \epsilon^{\rightarrow}_{I^3_1}(\text{Maria}_X, (\text{deu})_{I^3})))$$

(2-5b. — page 32) Maria diede il libro a Pietro

$$\text{a}^{\rightarrow}_{I^3_3}(\text{Pietro}_Z, \epsilon^{\leftarrow}_{I^3_2}((\text{il libro})_Y, \epsilon^{\rightarrow}_{I^3_1}(\text{Maria}_X, \text{diede}_{I^3})))$$

(2-5c. — page 32) Maria gli diede il libro

$$\epsilon^{\rightarrow}_{I^3_3}((\text{il libro})_Y, \epsilon^{\rightarrow}_{I^3_2}(\text{Maria}_X, \epsilon^{\leftarrow}_{I^3_1}(\text{gli}_Z, \text{diede}_{I^3})))$$

(2-5d. — page 33) Maria ~~gli~~ diede il libro

$$\epsilon^{\leftarrow}_{I^2_2}((\text{il libro})_Y, \epsilon^{\rightarrow}_{I^2_1}(\text{Maria}_X, \text{diede}_{I^2}))$$

(2-6a. — page 33) Der Mann gab der Frau das Auto

$$\epsilon^{\leftarrow}_{I^3_2[\text{Acc}]}((\text{das Auto})_Y, \epsilon^{\rightarrow}_{I^3_3[\text{Dat}]}((\text{der Frau})_Z, \epsilon^{\rightarrow}_{I^3_1[\text{Nom}]}((\text{der Mann})_X, \text{gab}_{I^3})))$$

(2-6b. — page 33) Der Mann gab es der Frau

$$\epsilon^{\rightarrow}_{I^3_3[\text{Dat}]}((\text{der Frau})_Z, \epsilon^{\leftarrow}_{I^3_2[\text{Acc}]}(\text{es}_Y, \epsilon^{\rightarrow}_{I^3_1[\text{Nom}]}((\text{der Mann})_X, \text{gab}_{I^3})))$$

(2-7a. — page 34) Wakonge wapele mbyia wanace

$$\epsilon^{\leftrightarrow}_{I^3_3}((\text{wanace})_Z, \epsilon^{\leftrightarrow}_{I^3_2}((\text{mbyia})_Y, \epsilon^{\leftrightarrow}_{I^3_1}((\text{Wakonge})_X, \text{wapele}_{I^3})))$$

(2-7b. — page 34) Wakonge wapele wanace mbyia

$$\epsilon^{\leftrightarrow}_{I^3_2}((\text{mbyia})_Y, \epsilon^{\leftrightarrow}_{I^3_3}((\text{wanace})_Z, \epsilon^{\leftrightarrow}_{I^3_1}((\text{Wakonge})_X, \text{wapele}_{I^3})))$$

A.3 Constructions

(3-1. — page 43) Ah!

$$\text{Ah!}_I$$

(3-2. — page 43) Dear me!

$$!^{\leftarrow}_{I^2_2}(\epsilon^{\rightarrow}_{O_2}((\cancel{\text{who am}}\ \text{dear})_A, \text{me}_{O_2}), \epsilon^{\leftarrow}_{I^2_1}(\cancel{(\text{you})}_{O_1}, \cancel{\text{consider}}_{I^2}))$$

$$!^{\leftarrow}_{I^2_2}(\epsilon^{\rightarrow}_{O_2}(\text{dear}_A, \text{me}_{O_2}), (\epsilon)_{I^2_1})$$

(3-3a. — page 46) Paul apologises to Liza.

$$\text{to}^{\leftarrow}_{I^2_2}(\text{Liza}_Y, \epsilon^{\rightarrow}_{I^2_1}(\text{Paul}_X, (\text{apologises})_{I^2}))$$

(3-3b. — page 46) Paul apologises.

$$\cancel{\text{to}}^{\leftarrow}_{I^2_2}(\cancel{\text{Liza}}_Y, \epsilon^{\rightarrow}_{I^2_1}(\text{Paul}_X, (\text{apologises})_{I^2}))$$

(3-4a. — page 46) Paul broke the vase.

$$\epsilon^{\leftarrow}_{I^2_2}((\text{the vase})_Y, \epsilon^{\rightarrow}_{I^2_1}(\text{Paul}_X, (\text{broke})_{I^2}))$$

(3-4b. — page 46) The vase broke.

$$\epsilon^{\leftarrow}_{I^2_2}((\text{the vase})_Y, \epsilon^{\rightarrow}_{I^2_1}(\cancel{\text{Paul}}_X, (\text{broke})_{I^2}))$$

(3-4c. — page 46) Paul bought a new vase.

$$\epsilon^{!}_{I^2_2}((\text{a new vase})_Y, \epsilon^{\rightarrow}_{I^2_1}(\text{Paul}_X, (\text{bought})_{I^2}))$$

(3-5a. — page 47) Ngarrkun-tu ari kanyirr.

$$\epsilon^{\rightarrow}_{I^2_2}(\text{kanyirr}_Y, \text{-tu}^{\rightarrow}_{I^2_1}(\text{Ngarrkun}_X, \text{ari}_{I^2}))$$

(3-5b. — page 48) Ngarrkun ari-li kanyirr-ku.

$$\text{-ku}^{\rightarrow}_{I_2^2 > E}(\text{kanyirr}_O, \epsilon^{\leftarrow}_{I_1^1}(\text{Ngarrkun}_X, (\text{ari-li})_{I^{2-1}}))$$

(3-5c. — page 48) Inú ga hasítte imásu.

$$\text{ga}^{\leftarrow}_{I_1^1}(\text{Inú}_X, \epsilon^{\rightarrow}_{I^1}(\text{imásu}_E, \text{hasítte}_{I^1}))$$

(3-5d. — page 48) Inú wa hasítte imásu.

$$\text{wa}^{\rightarrow}_{I_1^1}(\text{Inú}_X, \epsilon^{\rightarrow}_{I^1}(\text{imásu}_E, \text{hasítte}_{I^1}))$$

(3-5e. — page 49) Yukí wa sirói desu.

$$\epsilon^{\leftarrow}_{I_1^1}(\text{desu}_E, \text{wa}^{\rightarrow}_{I_1^1}(\text{Yukí}_X, \text{sirói}_A))$$

(3-5f. — page 49) Sonó bàra wa akái desu.

$$\epsilon^{\leftarrow}_{I_1^1}(\text{desu}_E, \text{wa}^{\rightarrow}_{I_1^1}(\epsilon^{\leftarrow}_X(\text{Sonó}_A, \text{barà}_X), \text{akái}_A))$$

(3-6a. — page 49) I sweat.

$$\epsilon^{\rightarrow}_{I_1^1}((\text{I}_X, (\text{sweat})_{I^1})$$

(3-6b. — page 49) Sa-laksha.

$$\epsilon^{\leftarrow}_{I_1^1}(\text{Sa-}_X, \text{laksha}_{I^1})$$

(3-6c. — page 50) (io) ho sudato.

$$\epsilon^{\rightarrow}_{I_1^2}((\text{io})_X, (\text{ho sudato})_{I^2})$$

(3-7a. — page 50) Paolo ha corso per un'ora.

$$\text{per}^{\leftarrow}_{E > I_2^2}(\epsilon^{\leftarrow}_O(\text{un}_A, \text{-a}^{\leftarrow}_O(\square_O, \text{or-}_O)),\ \epsilon^{\leftarrow}_{I_2^2}(\ \epsilon^{\rightarrow}_{A > Y_{aux}}(\text{cors-}^{\rightarrow}_A(\text{-o}_A, \epsilon^{\leftarrow}_{I_2^2}(\square Y,$$
$$\epsilon^{\rightarrow}_{I_1^2}((\cancel{\text{Paolo}})_X, \cancel{\text{corr}}_{I^2_{mean}}))), \square Y_{aux}),$$
$$\epsilon^{\rightarrow}_{I_1^2}(\text{-o}^{\leftarrow}_X(\square_X, \text{Paol-}_X), \text{ha}_{I^2})))$$

(3-7b. — page 51) Paolo è corso a casa.

$$\overleftarrow{\text{a}}_{E>I^2_2}(\overleftarrow{\text{-a}}_{O}(\square_O, \text{cas-}_O),\ \overrightarrow{\epsilon}_{I^2_2}(\overrightarrow{\epsilon}_{A>Y_{aux}}(\overrightarrow{\text{cors-}}_{A}(\text{-o}_A,$$
$$\overrightarrow{\epsilon}_{I^2_2}(\square_{Y_{mean}}, \overleftarrow{\epsilon}_{I^2_1}((\cancel{\text{Paolo}})_X, \cancel{\text{corr}}_{I^2_{mean}}))), \square_{Y_{aux}}),$$
$$\overleftarrow{\epsilon}_{I^2_1}(\overleftarrow{\text{-o}}_X(\square_X, \text{Paol-}_X), \text{è}_{I^2})))$$

(3-7c. — page 51) Paolo corse.

$$\overrightarrow{\epsilon}_{I^1_1}(\overleftarrow{\text{-o}}_X(\square_X, \text{Paol-}_X), \text{corse}_{I^2})$$

(3-8a. — page 51) ball table.

$$\overleftarrow{\epsilon}_{I^2_2}(\text{table}_E, \text{ball}_{I^2_1})$$

(3-8b. — page 51) (the) ball (is on the) table.

$$\cancel{\overrightarrow{\epsilon}}_{I^2_2}(\cancel{\overrightarrow{\epsilon}}_Y(\cancel{\overrightarrow{\text{on}}}_{E>A}(\square_D, (\cancel{\text{the}}\ \text{table})_O), (\cancel{\text{a ball}})_Y), \overleftarrow{\epsilon}_{I^2_1}((\cancel{\text{the}}\ \text{ball})_X, \cancel{\text{is}}_{I^2}))$$

(3-9a. — page 52) more milk.

$$\overleftarrow{\epsilon}_{I^2_2}(\text{more}_A, \text{milk}_Y)$$

(3-9b. — page 52) (I want) more milk.

$$\overleftarrow{\epsilon}_{I^2_2}(\overleftarrow{\epsilon}_Y(\text{more}_A, \text{milk}_Y), \cancel{\overrightarrow{\epsilon}}_{I^2_1}(\cancel{I}_X, \cancel{\text{want}}_{I^2}))$$

(3-10a. — page 52) more grapes.

$$\overleftarrow{\epsilon}_{I^2_2}(\text{more}_A, \text{grapes}_Y)$$

(3-10b. — page 52) (I want) more grapes.

$$\overleftarrow{\epsilon}_{I^2_2}(\overleftarrow{\epsilon}_Y(\text{more}_A, \text{grapes}_Y), \cancel{\overrightarrow{\epsilon}}_{I^2_1}(\cancel{I}_X, \cancel{\text{want}}_{I^2}))$$

(3-11a. — page 52) ancora mela.

$$\overleftarrow{\epsilon}_{I^2_2}(\text{ancora}_A, \text{mela}_Y)$$

(3-11b. — page 53) (io voglio) ancora mela.

$$\epsilon^{\leftarrow}_{I^2_2}(\epsilon^{\leftarrow}_{Y}(\text{ancora}_A, \text{mela}_Y), \cancel{\epsilon}^{\rightarrow}_{I^2_1}(\cancel{\text{io}}_X, \cancel{\text{voglio}}_{I^2}))$$

(3-12a. — page 52) ancora pasta.

$$\epsilon^{\leftarrow}_{I^2_2}(\text{ancora}_A, \text{pasta}_Y)$$

(3-12b. — page 53) (io voglio) ancora pasta.

$$\epsilon^{\leftarrow}_{I^2_2}(\epsilon^{\leftarrow}_{Y}(\text{ancora}_A, \text{pasta}_Y), \cancel{\epsilon}^{\rightarrow}_{I^2_1}(\cancel{\text{io}}_X, \cancel{\text{voglio}}_{I^2}))$$

(3-13a. — page 53) juice gone.

$$\epsilon^{\leftarrow}_{I^1}(\epsilon^{\rightarrow}_{X}(\text{gone}_A, \text{juice}_X), gov_{I^1})$$

(3-13b. — page 53) (the) juice (is) gone.

$$\epsilon^{\leftarrow}_{I^1_1}(\epsilon^{\rightarrow}_{X}((\text{gone})_{I>A}, \epsilon^{\leftarrow}_{X}(\cancel{\text{the}}_A, \text{juice}_X)), \cancel{\text{is}}_{I^1})$$

(3-14a. — page 53) o nofo pepe.

$$\epsilon^{\leftarrow}_{I^1_1}(\text{pepe}_X, \text{o nofo}_{I^1})$$

(3-14b. — page 53) o nofo ia pepe.

$$\text{ia}^{\rightarrow}_{I^1_1}(\text{pepe}_X, \text{o nofo}_{I^1})$$

(3-15a. — page 56) fortes ventos

$$\epsilon^{\leftarrow}_{O}((\text{fortes})_A, (\text{ventos})_O)$$

(3-15b. — page 56) ventos fortes

$$\epsilon^{\rightarrow}_{O}((\text{fortes})_A, (\text{ventos})_O)$$

(3-16a. — page 56) nostra memoria.

$$\text{-a}^{\leftarrow}_{O}((\text{nostr-})_A, (\text{memoria})_O)$$

(3-16b. — page 56) memoria nostri.

$$\text{-i}^{\rightarrow}_{O}((\text{nostr-})_A, (\text{memoria})_O)$$

(3-17a. — page 56) amor patris

$\text{-is}_{\mathrm{O}}^{\leftrightarrow}(\text{patr-}_{\mathrm{A}}, \text{amor}_{\mathrm{O}})$

(3-17b. — page 56) amor patris

$\text{-is}_{\mathrm{O}}^{\leftrightarrow}(\text{patr-}_{\mathrm{A}}, \text{amor}_{\mathrm{O}})$

(3-18a. — page 58) the elephant.

$\epsilon_{\mathrm{O}}^{\leftarrow}(\text{the}_{\mathrm{A}}, \text{elephant}_{\mathrm{O}})$

(3-18b. — page 58) that elephant.

$\epsilon_{\mathrm{O}}^{\rightarrow}(\text{that}_{\mathrm{A}}, \text{elephant}_{\mathrm{O}})$

(3-18c. — page 58) (I've seen) the one.

$\epsilon_{\mathrm{O}}^{\leftarrow}(\text{the}_{\mathrm{A}}, \text{one}_{\mathrm{O}})$

(3-18e. — page 58) (I've seen) that one.

$\epsilon_{\mathrm{O}}^{\rightarrow}(\text{that}_{\mathrm{A}}, \text{one}_{\mathrm{O}})$

(3-18f. — page 58) (I've seen) that.

$\epsilon_{\mathrm{O}}^{\rightarrow}(\text{that}_{\mathrm{A}}, \square_{\mathrm{O}})$

(3-19a. — page 60) Domani vengo.

$\epsilon_{\mathrm{I}}^{\leftarrow}(\text{domani}_{\mathrm{E}}, \epsilon_{\mathrm{I}_1^1}^{\rightarrow}((\text{io})_{\mathrm{X}}, \text{vengo}_{\mathrm{I}^1}))$

(3-19b. — page 60) Vengo domani.

$\epsilon_{\mathrm{I}}^{\leftarrow}(\text{domani}_{\mathrm{E}}, \epsilon_{\mathrm{I}_1^1}^{\rightarrow}((\text{io})_{\mathrm{X}}, \text{vengo}_{\mathrm{I}^1}))$

(3-19c. — page 62) Quando verrai?

$\epsilon_{\mathrm{I}_1^1}^{\rightarrow}((\text{quando})_{\mathrm{E}}, \epsilon_{\mathrm{I}_1^1}^{\rightarrow}((\text{tu})_{\mathrm{X}}, \text{verrai}_{\mathrm{I}^1}))$

(3-20. — page 63) that highly sensitive topic.

$\epsilon_{\mathrm{O}}^{\rightarrow}(\text{that}_{\mathrm{A}}, \epsilon_{\mathrm{O}}^{\leftarrow}(\epsilon_{\mathrm{A}}^{\rightarrow}(\text{-ly}_{\mathrm{E}}^{\rightarrow}(\square_{\mathrm{D}}, \text{high}_{\mathrm{A}}), (\text{sensitive})_{\ldots > \mathrm{A}}),$
$\text{topic}_{\mathrm{O}}))$

(3-21. — page 63) un vraiment beau paysage

$$\epsilon_{\mathrm{O}}^{\leftarrow}(\mathsf{un}_{\mathrm{A}}, \epsilon_{\mathrm{O}}^{\leftarrow}(\epsilon_{\mathrm{A}}^{\leftarrow}(\text{-}\mathsf{ment}_{\mathrm{E}}^{\rightarrow}(\Box_{\mathrm{D}}, \mathsf{vrai}_{\mathrm{A}}), \mathsf{beau}_{\mathrm{A}}), (\mathsf{paysage})_{\ldots > \mathrm{O}}))$$

A.4 Transformations

(4-1a. — page 73) Liza walk.

$$\epsilon_{\mathrm{I}_1^1}^{\rightarrow}(\mathsf{Liza}_{\mathrm{X}}, \mathsf{walk}_{\mathrm{I}^1})$$

(4-1b. — page 73) Liza walks.

$$\epsilon_{\mathrm{I}_1^1}^{\rightarrow}(\mathsf{Liza}_{\mathrm{X}}, (\mathsf{walks})_{\mathrm{I}^1})$$

(4-1c. — page 74) Liza [NOUN: PROPER, SINGULAR] walk-S [VERB: [TENSE: SIMPLE PRESENT: [3RD PERSON, SINGULAR]]].

$$\epsilon_{\mathrm{I}_1^1}^{\rightarrow}(\mathsf{Liza}_{\mathrm{X}:\ [\textsc{noun}];\ [\textsc{singular}]}, \epsilon_{\mathrm{I}^1}^{\leftarrow}(\text{-}\mathsf{s}_{\mathrm{I}:\ [3^{\mathrm{rd}}\ \textsc{person}];\ [\textsc{singular}]}, \mathsf{walk}_{\mathrm{I}^1}))$$

(4-2a. — page 75) Liza walks and Paul walks

$$\mathsf{and}_{\mathrm{X}}^{\leftrightarrow}(\epsilon_{\mathrm{I}_1^1}^{\rightarrow}(\mathsf{Paul}_{\mathrm{X}}, \epsilon_{\mathrm{I}^1}^{\leftarrow}(\text{-}\mathsf{s}_{\mathrm{I}}, \mathsf{walk}_{\mathrm{I}^1})), \epsilon_{\mathrm{I}_1^1}^{\rightarrow}(\mathsf{Liza}_{\mathrm{X}}, \epsilon_{\mathrm{I}^1}^{\leftarrow}(\text{-}\mathsf{s}_{\mathrm{I}}, \mathsf{walk}_{\mathrm{I}^1})))$$

(4-2b. — page 75) Liza and Paul walk.

$$\epsilon_{\mathrm{I}_1^1}^{\rightarrow}(\mathsf{and}_{\mathrm{X}}^{\leftrightarrow}(\mathsf{Paul}_{\mathrm{O}}, \mathsf{Liza}_{\mathrm{O}}), \mathsf{walk}_{\mathrm{I}^1})$$

(4-3a. — page 77) Liza is a quiet girl.

$$\epsilon_{\mathrm{I}_2^2}^{\rightarrow}(\epsilon_{\mathrm{Y}}^{\leftarrow}(\mathsf{a}_{\mathrm{A}}, \epsilon_{\mathrm{Y}}^{\rightarrow}(\mathsf{quiet}_{\mathrm{A}}, \mathsf{girl}_{\mathrm{Y}})), \epsilon_{\mathrm{I}_1^2}^{\leftarrow}(\mathsf{Liza}_{\mathrm{X}}, \mathsf{is}_{\mathrm{I}^2}))$$

(4-3b. — page 77) Liza is quiet.

$$\epsilon_{\mathrm{I}_2^2}^{\rightarrow}(\epsilon_{\mathrm{Y}}^{\rightarrow}(\mathsf{quiet}_{\mathrm{A}}, \Box_{\mathrm{Y}}), \epsilon_{\mathrm{I}_1^2}^{\leftarrow}(\mathsf{Liza}_{\mathrm{X}}, \mathsf{is}_{\mathrm{I}^2}))$$

(4-3c. — page 77) Be a quiet girl!

$$!_{\mathrm{I}_2^2}^{\rightarrow}(\epsilon_{\mathrm{Y}}^{\leftarrow}(\mathsf{a}_{\mathrm{A}}, \epsilon_{\mathrm{Y}}^{\rightarrow}(\mathsf{quiet}_{\mathrm{A}}, \mathsf{girl}_{\mathrm{Y}})), \epsilon_{\mathrm{I}_1^2}^{\leftarrow}(\Box_{\mathrm{X}}, \mathsf{be}_{\mathrm{I}^2}))$$

(4-3d. — page 77) Be quiet!

$$!_{\overrightarrow{I^2_2}}(\epsilon_{\overrightarrow{Y}}(\text{quiet}_{A}, \square_{Y}), \epsilon_{\overleftarrow{I^2_1}}(\square_{X}, \text{be}_{I^2}))$$

(4-4a. — page 77) Paul is studying maths.

$$\epsilon_{\overrightarrow{I^2_2}}(\text{-ing}_{\overrightarrow{Y}_{aux}}(\square_{D}, \epsilon_{\overleftarrow{I^2_2}}(\text{maths}_{Y_{mean}}, \epsilon_{\overrightarrow{I^2_1}}(\cancel{\text{Paul}}_{X_{mean}}, \text{study}_{I^2}))), \\ \epsilon_{\overleftarrow{I^2_1}}(\text{Paul}_{X_{aux}}, \text{is}_{I^2_{aux}}))$$

(4-4b. — page 77) Paul is going to study maths.

$$\epsilon_{\overrightarrow{I^2_2}}(\text{-ing}_{\overrightarrow{Y}_{aux}}(\square_{D}, \text{to}_{\overleftarrow{I^2_2}}(\epsilon_{\overleftarrow{Y}_{aux+}}(\square_{D}, \epsilon_{\overleftarrow{I^2_2}}(\text{maths}_{Y}, \epsilon_{\overrightarrow{I^2_1}}(\cancel{\text{Paul}}_{X}, \text{study}_{I^2}))), \\ \epsilon_{\overrightarrow{I^2_1}}(\cancel{\text{Paul}}_{X_{aux+}}, \text{go}_{I^2_{aux+}}))), \epsilon_{\overleftarrow{I^2_1}}(\text{Paul}_{X_{aux}}, \text{is}_{I^2_{aux}}))$$

(4-4c. — page 77) Maths is prepared by Paul.

$$\text{by}_{\overrightarrow{E>I}}(\text{Paul}_{X_{act}}, \epsilon_{\overrightarrow{I^2_2}}(\epsilon_{\overrightarrow{A>Y_{aux}}}(\text{-d}_{\overrightarrow{I^2_2>A}}(\cancel{\text{maths}}_{Y_{act}}, \\ \epsilon_{\overrightarrow{I^2_2}}(\cancel{\text{Paul}}_{X_{act}}, \text{prepare}_{I^2})), \\ \cancel{\text{maths}}_{Y_{act}}), \epsilon_{\overleftarrow{I^2_1}}(\text{maths}_{Y_{act} \rightsquigarrow X_{pass}}, \text{is}_{I^2_{aux}})))$$

(4-4d. — page 77) Maths is prepared.

$$\epsilon_{\overrightarrow{I^2_2}}(\epsilon_{\overrightarrow{A>Y_{aux}}}(\text{-d}_{\overrightarrow{I^2_2>A}}(\cancel{\text{maths}}_{Y_{act}}, \epsilon_{\overrightarrow{I^2_2}}(\square_{X_{act}}, \text{prepare}_{I^2})), \cancel{\text{maths}}_{Y_{act}}), \\ \epsilon_{\overleftarrow{I^2_1}}(\text{maths}_{Y_{act} \rightsquigarrow X_{pass}}, \text{is}_{I^2_{aux}}))$$

(4-4e. — page 77) Paul is prepared.

$$\epsilon_{\overrightarrow{I^2_2}}(\epsilon_{\overrightarrow{A>Y_{aux}}}(\text{-d}_{\overrightarrow{I^2_2>A}}(\square_{Y_{act}}, \epsilon_{\overrightarrow{I^2_2}}(\cancel{\text{Paul}}_{X_{act}}, \text{prepare}_{I^2})), \square_{Y_{act}}), \\ \epsilon_{\overleftarrow{I^2_1}}(\text{Paul}_{X_{act}}, \text{is}_{I^2_{aux}}))$$

(4-4f. — page 77) Be prepared!

$$\epsilon_{\overrightarrow{I^2_2}}(\epsilon_{\overrightarrow{A>Y_{aux}}}(\text{-d}_{\overrightarrow{I^2_2>A}}(\square_{Y_{act}}, \epsilon_{\overrightarrow{I^2_2}}(\cancel{\text{you}}_{X_{act}}, \text{prepare}_{I^2})), \square_{Y_{act}}), \\ \epsilon_{\overleftarrow{I^2_1}}(\cancel{\text{you}}_{X_{act}}, \text{be}_{I^2_{aux}}))$$

(4-5a. — page 84) The shah slept in the beds.

$$\text{in}_{\overrightarrow{I_2^2}}((\text{the beds})_{Y}, \epsilon_{\overrightarrow{I_1^2}}((\text{the shah})_{X}, \text{slept}_{I^2}))$$

(4-5b. — page 84) The beds were slept in by the shah.

$$\text{by}_{\overrightarrow{E>I}}((\text{the shah})_{X_{act}}, \epsilon_{\overrightarrow{I_2^2}}(\epsilon_{\overrightarrow{A>Y_{aux}}}(\text{slept in}_{\overrightarrow{I_2^2>A}}(\cancel{(\text{the beds})}_{Y_{act}},$$
$$\epsilon_{\overrightarrow{I_1^2}}(\cancel{(\text{the shah})}_{X_{act}}, \cancel{\text{sleep}}_{I^2_{mean}})), \square Y_{aux}),$$
$$\epsilon_{\overleftarrow{I_1^2}}((\text{the beds})_{Y_{act}}, \text{were}_{I^2_{aux}})))$$

(4-5c. — page 88) Shah's sleeping.

$$\text{'s}_{\overrightarrow{A>O}}(\text{Shah}_{O}, \text{-ing}_{\overrightarrow{O}}(\square_{D}, \text{sleep}_{I}))$$

(4-6a. — page 84) A camel stepped on the package.

$$\text{-ped on}_{\overrightarrow{I_2^2}}(\cancel{(\text{the package})}_{Y}, \epsilon_{\overrightarrow{I_1^2}}((\text{A camel})_{X}, \text{step}_{I^2}))$$

(4-6b. — page 84) The package was stepped on by a camel.

$$\text{by}_{\overrightarrow{E>I}}((\text{a camel})_{X_{act}}, \epsilon_{\overrightarrow{I_2^2}}(\epsilon_{\overrightarrow{A>Y_{aux}}}(\text{-ped on}_{\overrightarrow{I_2^2>A}}(\cancel{(\text{the package})}_{Y_{act}},$$
$$\epsilon_{\overrightarrow{I_1^2}}(\cancel{(\text{the camel})}_{X_{act}}, \text{step}_{I^2_{mean}})), \square Y_{aux}),$$
$$\epsilon_{\overleftarrow{I_1^2}}((\text{the package})_{Y_{act}}, \text{was}_{I^2_{aux}})))$$

(4-6c. — page 88) The camel's stepping.

$$\text{'s}_{\overrightarrow{O}}((\text{The camel})_{O}, \text{-p-ing}_{\overrightarrow{O}}(\square_{D}, \text{step}_{I}))$$

$$\epsilon_{\overleftarrow{O}}(\text{The}, \text{'s}_{\overrightarrow{O}}(\text{camel}_{O}, \text{-p-ing}_{\overrightarrow{O}}(\square_{D}, \text{step}_{I})))$$

(4-7a. — page 84) The ice cube melted in the oven.

$$\text{in}_{\overleftarrow{I_2^2}}(\epsilon_{\overrightarrow{O}}(\text{the}_{A}, \text{oven}_{O}), \epsilon_{\overleftarrow{I_1^2}}(\epsilon_{\overrightarrow{O}}(\text{The}_{A}, \epsilon_{\overrightarrow{O}}(\text{ice}_{O>A}, \text{cube}_{O})),$$
$$(\text{melted})_{I^2}))$$

(4-7c. — page 88) The melting of the ice cube.

$$\text{of}_{\overleftarrow{O}}(\epsilon_{\overrightarrow{O}}(\text{the}_{A}, \epsilon_{\overrightarrow{A>O}}(\text{ice}_{O}, \text{cube}_{O})), \epsilon_{\overleftarrow{O}}(\text{the}_{A}, \text{-ing}_{\overrightarrow{O}}(\square_{D}, \text{melt}_{I})))$$

(4-8a. — page 84) The toothpaste oozed into the sink.

$$\mathsf{in}^{\leftarrow}_{I^2_2}(\epsilon^{\leftarrow}_{O}(\mathsf{the}_A, \mathsf{sink}_O), \epsilon^{\leftarrow}_{I^2_1}(\epsilon^{\leftarrow}_{O}(\mathsf{The}_A, (\mathsf{toothpaste})_O), (\mathsf{oozed})_{I^2}))$$

(4-8c. — page 88) The ooze of the toothpaste.

$$\mathsf{of}^{\rightarrow}_{O}(\epsilon^{\rightarrow}_{O}(\mathsf{the}_A, (\mathsf{toothpaste})_O), \epsilon^{\leftarrow}_{O}(\mathsf{The}_A, \epsilon^{\rightarrow}_{O}(\square_D, \mathsf{ooze}_I)))$$

(4-9a. — page 84) Little Red Riding Hood vanished in the woods.

$$\mathsf{in}^{\leftarrow}_{I}(\epsilon^{\leftarrow}_{O}(\mathsf{the}_A, (\mathsf{woods})_O), \epsilon^{\rightarrow}_{I^1_1}((\mathsf{Little\ Red\ Riding\ Hood})_O, (\mathsf{vanished})_{I^1}))$$

(4-9c. — page 88) The vanishing of Little Red Riding Hood.

$$\mathsf{of}^{\leftarrow}_{O}((\mathsf{Little\ Red\ Riding\ Hood})_O, \epsilon^{\leftarrow}_{O}(\mathsf{the}_A, \mathsf{\text{-}ing}^{\leftarrow}_{O}(\square_D, \mathsf{vanish}_I)))$$

(4-10a. — page 89) It's a car.

$$\epsilon^{\leftarrow}_{I^2_2}(\epsilon^{\leftarrow}_{O}(\mathsf{a}_A, \mathsf{car}_O), \mathsf{'}^{\rightarrow}_{I^2_1}(\mathsf{It}_O, \mathsf{is}_{I^2}))$$

(4-10b. — page 89) That's the elephant.

$$\epsilon^{\leftarrow}_{I^2_2}(\epsilon^{\leftarrow}_{O}(\mathsf{the}_A, \mathsf{elephant}_O), \mathsf{'}^{\rightarrow}_{I^2_1}(\mathsf{that}_O, \mathsf{is}_{I^2}))$$

(4-11a. — page 89) That's my mum.

$$\epsilon^{\rightarrow}_{I^2_2}(\epsilon^{\rightarrow}_{O}(\mathsf{my}_A, \mathsf{mum}_O), \mathsf{'}^{\leftarrow}_{I^2_1}(\mathsf{that}_O, \mathsf{is}_{I^2}))$$

(4-11b. — page 89) It's Paul's brother.

$$\epsilon^{\rightarrow}_{I^2_2}(\mathsf{'s}^{\rightarrow}_{O}(\mathsf{Paul}_O, \mathsf{brother}_O), \mathsf{'}^{\leftarrow}_{I^2_1}(\mathsf{It}_O, \mathsf{is}_{I^2}))$$

(4-11c. — page 90) There are my shoes.

$$\epsilon^{\rightarrow}_{I^2_2}((\mathsf{my\ shoes})_Y, \epsilon^{\leftarrow}_{I^2_1}(\square_X, \mathsf{th\text{-}}^{\leftarrow}_{E > I^2}(\mathsf{\text{-}ere}_E, \mathsf{are}_I)))$$

(4-11d. — page 90) There is Liza.

$$\epsilon^{\rightarrow}_{I^2_2}((\mathsf{Liza})_Y, \epsilon^{\leftarrow}_{I^2_1}(\square_X, \mathsf{th\text{-}}^{\leftarrow}_{E > I^2}(\mathsf{\text{-}ere}_E, \mathsf{is}_I)))$$

(4-11e. — page 90) Here's a gift.

$\epsilon^{\rightarrow}_{I^2_2}((\text{a gift})_Y, \epsilon^{\leftarrow}_{I^2_1}(\square_X, \text{h-}^{\leftarrow}_{E>I^2}(\text{-ere}_E, \text{is}_I)))$

(4-12a. — page 90) My shoes are there.

$\text{th-}^{\rightarrow}_{E>I^2_2}(\text{-ere}_E, \epsilon^{\rightarrow}_{I^2_2}(\square_Y, \epsilon^{\leftarrow}_{I^2_1}((\text{My shoes})_X, \text{are}_{I^2})))$

(4-12b. — page 90) Liza is there.

$\text{th-}^{\rightarrow}_{E>I^2_2}(\text{-ere}_E, \epsilon^{\rightarrow}_{I^2_2}(\square_Y, \epsilon^{\leftarrow}_{I^2_1}((\text{Liza})_X, \text{is}_{I^2})))$

(4-12c. — page 90) Paul is here.

$\text{h-}^{\rightarrow}_{E>I^2_2}(\text{-ere}_E, \epsilon^{\rightarrow}_{I^2_2}(\square_Y, \epsilon^{\leftarrow}_{I^2_1}(\text{Paul}_X, \text{is}_{I^2})))$

(4-13. — page 91) kedim var.

$\epsilon^{\rightarrow}_{I^2_2}(\text{-m}_Y, \epsilon^{\leftarrow}_{I^2_1}(\text{kedi}_X, \text{var}_{I^2}))$

(4-14a. — page 91) Paul has a Bentley.

$\epsilon^{\leftarrow}_{I^2_2}((\text{a Bentley})_O, \epsilon^{\rightarrow}_{I^2_1}(\text{Paul}_O, \text{has}_{I^2}))$

(4-14b. — page 91) Paul has got a Bentley.

$$\epsilon^{\leftarrow}_{I^2_2}(\epsilon^{\rightarrow}_{Y_{aux}}(\epsilon^{\rightarrow}_{I^2_2>A}((\text{a Bentley})_{Y_{mean}}, \epsilon^{\rightarrow}_{I^2_1}(\text{\sout{Paul}}_{X_{mean}}, \text{got}_{I^2_{mean}})), \square_{Y_{aux}}), \epsilon^{\rightarrow}_{I^2_1}(\text{Paul}_{X_{aux}}, \text{has}_{I^2_{aux}}))$$

(4-14c. — page 91) Paul has got a Ford.

$$\epsilon^{\leftarrow}_{I^2_2}(\epsilon^{\rightarrow}_{Y_{aux}}(\epsilon^{\rightarrow}_{I^2_2>A}((\text{a Ford})_{Y_{mean}}, \epsilon^{\rightarrow}_{I^2_1}(\text{\sout{Paul}}_{X_{mean}}, \text{got}_{I^2_{mean}})), \square_{Y_{aux}}), \epsilon^{\rightarrow}_{I^2_1}(\text{Paul}_{X_{aux}}, \text{has}_{I^2_{aux}}))$$

(4-14d. — page 91) Paul has washed his car.

$$\epsilon^{\leftarrow}_{I^2_2}(\,\epsilon^{\rightarrow}_{Y_{aux}}(\text{-ed}^{\rightarrow}_{I^2_2>A}((\text{his car})_{Y_{mean}},\ \epsilon^{\rightarrow}_{I^2_1}(\sout{\text{Paul}}_{X_{mean}},\ \text{wash}_{I^2_{mean}})),\ \square Y_{aux}),\ \epsilon^{\rightarrow}_{I^2_1}(\text{Paul}_{X_{aux}},\ \text{has}_{I^2_{aux}}))$$

(4-14e. — page 93) Paul got a Ford.

$$\epsilon^{\leftarrow}_{I^2_2}((\text{a Ford})_O,\ \epsilon^{\rightarrow}_{I^2_1}(\text{Paul}_O,\ \text{got}_{I^2}))$$

(4-14f. — page 93) Paul washed his car.

$$\epsilon^{\leftarrow}_{I^2_2}((\text{his car})_O,\ \epsilon^{\rightarrow}_{I^2_1}(\text{Paul}_O,\ (\text{washed})_{I^2}))$$

(4-15a. — page 93) Paola ha corso la maratona entro quattro ore.

$$\text{entro}^{\leftarrow}_{E>I^2_2}(\epsilon^{\leftarrow}_{O}(\text{quattro}_A,\ \text{-e}^{\leftarrow}_{O}(\square O,\ \text{or-}_O)),\ \epsilon^{\leftarrow}_{I^2_2}(\,\epsilon^{\rightarrow}_{Y_{aux}}(\text{cors-}^{\rightarrow}_{A}(\text{-o}_A,\ \epsilon^{\leftarrow}_{I^2_2}((\text{la maratona})_Y,\ \epsilon^{\rightarrow}_{I^2_1}((\sout{\text{Paola}})_X,\ \text{cor}\sout{\text{-re}}_{I^2_{mean}}))),\ \square Y_{aux})\,,\ \epsilon^{\rightarrow}_{I^2_1}(\text{-a}^{\leftarrow}_{X}(\square X,\ \text{Paol-}_X),\ \text{ha}_{I^2})))$$

(4-15b. — page 93) Paola è corsa a casa.

$$\text{a}^{\leftarrow}_{E>I^2_2}(\text{-a}^{\leftarrow}_{O}(\square O,\ \text{cas-}_O),\ \ \epsilon^{\rightarrow}_{I^2_2}(\,\epsilon^{\rightarrow}_{Y_{aux}}(\text{-s-}^{\rightarrow}_{A}(\text{-a}_A,\ \epsilon^{\rightarrow}_{I^2_2}(\square Y_{mean},\ \epsilon^{\leftarrow}_{I^2_1}((\sout{\text{Paola}})_X,\ \text{cor}\sout{\text{-re}}_{I^2_{mean}}))),\ \square Y_{aux}),\ \epsilon^{\leftarrow}_{I^2_1}(\text{-a}^{\leftarrow}_{X}(\square X,\ \text{Paol-}_X),\ \text{è}_{I^2})))$$

(4-16a. — page 96) Mario ha lavato la camicia.

$$\epsilon^{\leftarrow}_{I^2_2}(\,\epsilon^{\leftarrow}_{Y_{aux}}((\text{lavat-}\mathbf{o}\text{ la camicia})_{I^2>A},\ \square Y_{aux})\,,\ \epsilon^{\rightarrow}_{I^2_1}((\text{Paolo})_X,\ \text{ha}_{I^2}))$$

(4-16b. — page 96) Mario ha la camicia lavata.

$$\epsilon^{\leftarrow}_{I^2_2}(\,\epsilon^{\rightarrow}_{Y_{aux}}((\text{lavat-}\mathbf{a})_{I^2>A},\ (\text{la camicia})_{Y_{aux}})\,,\ \epsilon^{\rightarrow}_{I^2_1}((\text{Paolo})_X,\ \text{ha}_{I^2}))$$

(4-17. — page 98) Meeting people is easy.

$$\epsilon_{\overrightarrow{I_2^2}}(\epsilon_{\overrightarrow{Y_{aux}}}(\text{easy}_{A}, \cancel{\text{something}}_{Y_{be}}),\ \epsilon_{\overleftarrow{I_1^2}}(\text{-ing}_{\overrightarrow{X_{be}}}(\Box_{D}, \epsilon_{\overleftarrow{I_2^2}}(\text{people}_{Y_{meet}}, \epsilon_{\overrightarrow{I_1^2}}(\Box_{X_{meet}}, \text{meet}_{I^2_{meet}}))), \text{is}_{I^2_{be}}))$$

(4-18a. — page 99) The machine needs fixing.

$$\epsilon_{\overleftarrow{I_2^2}}(\text{-ing}_{\overrightarrow{I_2^2} > Y_{need}}((\cancel{\text{the machine}})_{Y_{fix}}, \epsilon_{\overrightarrow{I_1^2}}((\cancel{\text{somebody}})_{X_{fix}}, \text{fix}_{I^2_{fix}})), \epsilon_{\overrightarrow{I_1^2}}((\text{the machine})_{X_{need}}, (\text{needs})_{I^2_{need}}))$$

(4-18b. — page 99) My house requires painting.

$$\epsilon_{\overleftarrow{I_2^2}}(\text{-ing}_{\overrightarrow{I_2^2} > Y_{req}}((\cancel{\text{my house}})_{Y_{paint}}, \epsilon_{\overrightarrow{I_1^2}}((\cancel{\text{somebody}})_{X_{paint}}, \text{paint}_{I^2_{paint}})), \epsilon_{\overrightarrow{I_1^2}}((\text{My house})_{X_{req}}, (\text{requires})_{I^2_{req}}))$$

(4-19. — page 99) Paul is a working class hero.

$$\epsilon_{\overrightarrow{I_2^2}}(\text{Paul}_{O_1}, \epsilon_{\overleftarrow{I_1^2}}(\epsilon_{\overleftarrow{Y_{be}}}(\text{a}_{A}, \epsilon_{\overleftarrow{A} > Y_{be}}(\epsilon_{\overleftarrow{O}}(\text{-ing}_{\overrightarrow{I_2^2} > O}((\cancel{\text{a work}})_{Y_{work}}, \epsilon_{\overrightarrow{I_1^2}}((\cancel{\text{worker}})_{X_{work}}, \text{work}_{I^2_{work}})), \text{class}_{O}), \text{hero}_{Y_{be}})), (\text{is})_{I^2}))$$

(4-20a. — page 100) Paul is a dead man walking.

$$\epsilon_{\overrightarrow{I_2^2}}(\text{Paul}_{O_1}, \epsilon_{\overleftarrow{I_1^2}}(\epsilon_{\overleftarrow{Y_{be}}}(\text{a}_{A}, \cancel{\text{wh-}}_{\overrightarrow{A}}(\text{ -ing}_{\overrightarrow{O}}(\cancel{\text{-o}}_{X_{walk}}, \text{walk}_{I^1_{walk}}), \epsilon_{\overrightarrow{Y_{be}}}(\text{dead}_{I > A}, \text{man}_{Y_{be}}))), (\text{is})_{I^2}))$$

(4-21. — page 102) Without telling him, Liza cannot be happy.

$$\text{without}_{\overleftarrow{E} > I}(\text{-ing}_{\overrightarrow{I_3^3} > O}(\text{him}_{Z}, \epsilon_{\overleftarrow{I_2^3}}(\cancel{\text{something}}_{Y}, \epsilon_{\overleftarrow{I_1^3}}(\cancel{\text{Liza}}_{X}, \text{tell}_{I_3^3}))), (\ldots)_{I})$$

(4-22a. — page 103) Avendo telefonato Lisa, Paolo pianse.

$$\text{-ndo}_{\overrightarrow{I_2^2} > E > I}(\text{-e}_{\overleftarrow{I_2^2}}(\ \epsilon_{\overrightarrow{Y_{aux}}}(\cancel{\text{a}}\rightsquigarrow\text{-ato}_{\overrightarrow{I_2^2} > A}(\cancel{\text{Paolo}}_{Y}, \epsilon_{\overrightarrow{I_1^2}}(\text{Lisa}_{X_{mean}}, \text{telefon}_{I^2_{mean}})), \Box_{Y_{aux}}), \epsilon_{\overrightarrow{I_1^2}}(\cancel{\text{Lisa}}_{X_{aux}}, \cancel{\text{ha}}\rightsquigarrow\text{av-}_{I^2_{aux}})), (\text{Paolo pianse})_{I})$$

(4-22c. — page 103) Essendo partita Lisa, Paolo pianse.

$$\text{-ndo}_{\overrightarrow{I^2_2}} > E > I^{(\text{-e}_{\overrightarrow{I^2_2}}(\epsilon_{\overrightarrow{Y_{aux}}}(\text{-it-}_{\overrightarrow{I^1_1}} > A^{(\text{-a}_A,\, \epsilon_{\overrightarrow{I^1_1}}(\text{Lisa}_{X_{mean}},\, \text{part}_{I^1_{mean}})),\, \square Y_{aux}),\, \epsilon_{\overleftarrow{I^2_1}}(\cancel{\text{Lisa}}_{X_{aux}},\, \text{ess}_{I^2_{aux}})),\, (\text{Paolo pianse})_I)$$

(4-22d. — page 103) Partita Lisa, Paolo pianse.

$$\cancel{\text{-ndo}}_{\overrightarrow{I^2_2}} > E > I^{(\cancel{\text{-e}}_{\overrightarrow{I^2_2}}(\epsilon_{\overrightarrow{Y_{aux}}}(\text{-it-}_{\overrightarrow{I^1_1}} > A^{(\text{-a}_A,\, \epsilon_{\overrightarrow{I^1_1}}(\text{Lisa}_{X_{mean}},\, \text{part}_{I^1_{mean}})),\, \square Y_{aux}),\, \epsilon_{\overleftarrow{I^2_1}}(\cancel{\text{Lisa}}_{X_{aux}},\, \cancel{\text{ess}}_{I^2_{aux}})),\, (\text{Paolo pianse})_I)$$

(4-23a. — page 106) A Lisa piace ballare.

$$\text{a}_{\overrightarrow{I^2_2}}(\text{Lisa}_Y,\, \epsilon_{\overleftarrow{I^2_1}}(\text{-are}_{\overrightarrow{X_{piace}}}(\square Y,\, \epsilon_{\overrightarrow{I^2_1}}(\cancel{\text{Lisa}}_X,\, \text{ball}_{I^2})),\, \text{piace}_{I^2}))$$

(4-24. — page 106) Paolo vuole provare ad andare a fare una vacanza.

$$\epsilon_{\overleftarrow{I^2_2}}(\text{-are}_{\overrightarrow{I^2_2}} > Y^{(\text{ad -are}_{\overrightarrow{I^2_2}} > Y^{(\text{a -are}_{\overrightarrow{I^2_2}} > Y^{((\text{una vacanza})_Y,}}}$$
$$\epsilon_{\overrightarrow{I^2_1}}(\text{Paolo}_X,\, \text{f}_{I^2})),\, \epsilon_{\overrightarrow{I^2_1}}(\cancel{\text{Paolo}}_X,\, \text{and}_{I^2})),\, \epsilon_{\overrightarrow{I^2_1}}(\cancel{\text{Paolo}}_X,\, \text{prov}_{I^2})),$$
$$\epsilon_{\overrightarrow{I^2_1}}(\text{Paolo}_X,\, \text{vuole}_{I^2}))$$

(4-25. — page 106) Lamento eles ter-em perdido o comboio.

$$\epsilon_{\overrightarrow{I^2_2}}(\text{-em}_{\overrightarrow{I^2}} > Y_{lam.}{}^{(\epsilon_{\overrightarrow{Y_{ter}}}(\text{-ido}_{\overrightarrow{I^2_2}} > A^{((\text{o comboio})_{Y_{perd}},}}$$
$$\epsilon_{\overrightarrow{I^2_1}}(\cancel{\text{eles}}_{X_{perd}},\, \text{perd}_{I^2_{perd}})),\, \square Y_{ter}),\, \epsilon_{\overrightarrow{I^2_1}}(\text{eles}_{X_{ter}},\, \text{ter}_{I^2_{ter}})),\, (\text{lamento})_{I^2_1})$$

(4-26a. — page 107) Mario ha una camicia lavabile a secco.

$$\epsilon_{\overleftarrow{I^2_2}}(\epsilon_{\overrightarrow{Y}}(\text{a}_{\overrightarrow{A}}((\text{secco})_I > A,\, \text{-bil}_{\overrightarrow{A}}(\text{-e}_A,\, \epsilon_{\overrightarrow{I^2_2}}(\text{-a}_I,\, \epsilon_{\overrightarrow{I^2_2}}((\cancel{\text{una camicia}})_Y,$$
$$\epsilon_{\overrightarrow{I^2_1}}(\text{Mario}_X,\, \text{lav}_{I^2}))))),\, (\text{una camicia})_Y),\, \epsilon_{\overrightarrow{I^2_1}}(\text{Mario}_X,\, \text{ha}_{I^2}))$$

(4-27. — page 111) Le livre de Pierre.

$$\epsilon_{\overrightarrow{O}}(\text{de}_{\overrightarrow{A}}(\square_D,\, \text{Pierre}_O),\, (\text{le livre})_O)$$

(4-28. — page 111) Une tasse de thé.

$$\epsilon_{\overrightarrow{O}}(\text{de}_{\overrightarrow{A}}(\square_D,\, \text{thé}_O),\, (\text{une tasse})_O)$$

(4-29. — page 111) Une salle de classe.

$$\epsilon^{\rightarrow}_{O}(\text{de}^{\rightarrow}_{A}(\square_D, \text{classe}_O), (\text{une salle})_O)$$

(4-30. — page 112) Je écrire de la main gauche.

$$\text{de}^{\leftarrow}_{E}((\text{la main gauche})_O, (\text{je écrire})_I)$$

(4-31. — page 112) Je répète de mémoire.

$$\text{de}^{\leftarrow}_{E}((\text{mémoire})_O, (\text{je répète})_I)$$

(4-32. — page 112) Le train (parti) de Paris…

$$\epsilon^{\rightarrow}_{O}(\text{de}^{\leftarrow}_{E}(\text{Paris}_O, \cancel{(\text{parti})}_{I>A}), (\text{le train})_O)$$

(4-33. — page 113) Liza wiped the table clean.

$$\epsilon^{\leftarrow}_{I^2_2}(\cancel{\text{wh-}}^{\rightarrow}_{Y}(\cancel{\text{-ed}}^{\rightarrow}_{A}(\square_D, \epsilon^{\rightarrow}_{I^2_2}(\cancel{\text{-ich}}_Y, \epsilon^{\rightarrow}_{I^2_1}(\cancel{\text{she}}_X, \text{clean}_{I^2}))), (\text{the table})_Y), \epsilon^{\rightarrow}_{I^2_1}(\text{Liza}_X, (\text{wiped})_{I^2}))$$

(4-34a. — page 113) Paul has got his car washed.

$$\epsilon^{\leftarrow}_{I^2_2}(\epsilon^{\rightarrow}_{Y_{aux}}(\square Y_{aux}, \epsilon^{\rightarrow}_{I^2_2>A}(\epsilon^{\rightarrow}_{I^2_1}(\cancel{\text{Paul}}_{X_{got}}, \text{got}_{I^2_{got}}), \cancel{\text{wh-}}^{\rightarrow}_{Y_{got}}(\text{-ed}^{\rightarrow}_{A}(\square_D, \epsilon^{\rightarrow}_{I^2_2}(\cancel{\text{-ich}}_Y, \epsilon^{\rightarrow}_{I^2_1}(\text{he}_X, \text{wash}_{I^2}))), (\text{his car})_{Y_{got}}))), \epsilon^{\rightarrow}_{I^2_1}(\text{Paul}_{X_{aux}}, \text{has}_{I^2_{aux}}))$$

(4-34b. — page 115) He drove her mad.

$$\epsilon^{\leftarrow}_{I^2_2}(\epsilon^{\rightarrow}_{I^2_1}(\text{He}_{X_{mean}}, \text{drove}_{I^2_{mean}}), \cancel{\text{wh-}}^{\rightarrow}_{Y_{mean}}(\text{her}_{Y_{mean}}, \cancel{\text{-ed}}^{\rightarrow}_{A}(\square_D, \epsilon^{\leftarrow}_{I^2_2}(\text{mad}_{A>Y}, \epsilon^{\leftarrow}_{I^2_1}(\cancel{\text{she}}_X, \cancel{\text{is}}_{I^2})))))$$

(4-35a. — page 115) Alfred can pay because he is rich.

$$\text{because}^{\leftarrow}_{I}(\epsilon^{\rightarrow}_{I^2_2}(\epsilon^{\rightarrow}_{Y}(\text{rich}_A, \square_Y), \epsilon^{\rightarrow}_{I^2_1}(\text{he}_X, \text{is}_{I^2_{dep}})), \epsilon^{\leftarrow}_{I^2_2}(\text{pay}_{I>O}, \epsilon^{\rightarrow}_{I^2_1}(\text{Alfred}_X, \text{can}_{I^2_{gov}})))$$

(4-35a. — page 115) Alfred peut payer parce que il est riche.

$$\text{parce que}_{\overleftarrow{I}}(\epsilon_{\overrightarrow{I^2_2}}(\epsilon_{\overrightarrow{Y}}(\text{riche}_A, \square_Y), \epsilon_{\overrightarrow{I^2_1}}(\text{il}_X, \text{est}_{I^2_{dep}})), \epsilon_{\overleftarrow{I^2_2}}(\text{payer}_{I>O}, \epsilon_{\overrightarrow{I^2_1}}(\text{Alfred}_X, \text{puet}_{I^2_{gov}})))$$

(4-35a. — page 115) Alfred può pagare perché è ricco.

$$\text{perché}_{\overleftarrow{I}}(\epsilon_{\overrightarrow{I^2_2}}(\epsilon_{\overrightarrow{Y}}(\text{ricco}_A, \square_Y), \epsilon_{\overrightarrow{I^2_1}}(\text{egli}_X, \text{è}_{I^2_{dep}})), \epsilon_{\overleftarrow{I^2_2}}(\text{pagare}_{I>O}, \epsilon_{\overrightarrow{I^2_1}}(\text{Alfred}_X, \text{può}_{I^2_{gov}})))$$

(4-35b. — page 115) Alfred can pay hence he is rich.

$$\text{hence}_{\overrightarrow{I}}(\epsilon_{\overrightarrow{I^2_2}}(\epsilon_{\overrightarrow{Y}}(\text{rich}_A, \square_Y), \epsilon_{\overrightarrow{I^2_1}}(\text{he}_X, \text{is}_{I^2_{dep}})), \epsilon_{\overleftarrow{I^2_2}}(\text{pay}_{I>O}, \epsilon_{\overrightarrow{I^2_1}}(\text{Alfred}_X, \text{can}_{I^2_{gov}})))$$

(4-35b. — page 115) Alfred peut payer donc il est riche.

$$\text{donc}_{\overrightarrow{I}}(\epsilon_{\overrightarrow{I^2_2}}(\epsilon_{\overrightarrow{Y}}(\text{riche}_A, \square_Y), \epsilon_{\overrightarrow{I^2_1}}(\text{il}_X, \text{est}_{I^2_{dep}})), \epsilon_{\overleftarrow{I^2_2}}(\text{payer}_{I>O}, \epsilon_{\overrightarrow{I^2_1}}(\text{Alfred}_X, \text{puet}_{I^2_{gov}})))$$

(4-35b. — page 115) Alfred può pagare dunque è ricco.

$$\text{dunque}_{\overrightarrow{I}}(\epsilon_{\overrightarrow{I^2_2}}(\epsilon_{\overrightarrow{Y}}(\text{ricco}_A, \square_Y), \epsilon_{\overrightarrow{I^2_1}}(\text{egli}_X, \text{è}_{I^2_{dep}})), \epsilon_{\overleftarrow{I^2_2}}(\text{pagare}_{I>O}, \epsilon_{\overrightarrow{I^2_1}}(\text{Alfred}_X, \text{può}_{I^2_{gov}})))$$

(4-36a. — page 116) Alfred, who is rich, can pay.

$$\epsilon_{\overleftarrow{I^2_2}}(\text{pay}_{I>Y}, \epsilon_{\overrightarrow{I^2_1}}(\text{wh-}_{\overleftarrow{I}>A>X}(\epsilon_{\overrightarrow{I^2_2}}(\text{rich}_{A>Y_{dep}}, \epsilon_{\overrightarrow{I^2_1}}(\text{-o}_{X_{dep}}, \text{is}_{I^2_{dep}})), \text{Alfred}_X), \text{can}_{I^2}))$$

(4-36b. — page 116) Alfred, who can pay, is rich.

$$\epsilon_{\overleftarrow{I^2_2}}(\text{rich}_{A>Y}, \epsilon_{\overrightarrow{I^2_1}}(\text{wh-}_{\overleftarrow{I}>A>X}(\epsilon_{\overrightarrow{I^2_2}}(\text{pay}_{I>Y_{dep}}, \epsilon_{\overrightarrow{I^2_1}}(\text{-o}_{X_{dep}}, \text{can}_{I^2_{dep}})), \text{Alfred}_X), \text{is}_{I^2}))$$

(4-37a. — page 120) the man who bought the book for the girl.

$$\text{wh-}_{\overrightarrow{I > A > O}}(\text{for}_{\overrightarrow{E > I}}((\text{the girl})_O, \epsilon_{\overleftarrow{I^4_2}}((\text{the book})_Y, \epsilon_{\overrightarrow{I^4_1}}(\text{-o}_X, (\text{bought})_{I^4_{dep}}))), (\text{the man})_O)$$

(4-37b. — page 120) the book which the man bought for the girl.

$$\text{wh-}_{\overrightarrow{I > A > O}}(\text{for}_{\overrightarrow{E > I}}((\text{the girl})_O, \epsilon_{\overleftarrow{I^4_2}}((\text{-ich})_Y, \epsilon_{\overrightarrow{I^4_1}}((\text{the man})_X, (\text{bought})_{I^4_{dep}}))), (\text{the book})_O)$$

(4-37c. — page 120) the girl for whom the man bought the book.

$$\text{wh-}_{\overrightarrow{I > A > O}}(\text{for}_{\overrightarrow{E > I}}((\text{-om})_O, \epsilon_{\overleftarrow{I^4_2}}((\text{the book})_Y, \epsilon_{\overrightarrow{I^4_1}}(\text{the man}_X, (\text{bought})_{I^4_{dep}}))), (\text{the girl})_O)$$

(4-37d. — page 120) the boy whose book the man bought for the girl.

$$\text{wh-}_{\overrightarrow{I > A > O}}(\text{for}_{\overrightarrow{E > I}}((\text{the girl})_O, \epsilon_{\overleftarrow{I^4_2}}(\epsilon_{\overrightarrow{Y}}(\text{-ose}_A, \text{book}_Y), \epsilon_{\overrightarrow{I^4_1}}((\text{the man})_X, (\text{bought})_{I^4_{dep}}))), (\text{the boy})_O)$$

(4-37f. — page 125) Who bought the book for the girl? The man.

$$?_{\overrightarrow{I}}((\text{the man})_O, \text{wh-}_{\overrightarrow{I > A > O}}(\text{for}_{\overrightarrow{E^4_2}}((\text{the girl})_O, \epsilon_{\overleftarrow{I^4_2}}((\text{the book})_Y, \epsilon_{\overrightarrow{I^4_1}}(\text{-o}_X, (\text{bought})_{I^4}))), \square_O))$$

(4-38a. — page 121) When Paul will come back Liza will smile.

$$\text{wh-}_{\overrightarrow{E > I^1_1}}(\epsilon_{\overleftarrow{I^1_1}}(\text{-en}_E, \epsilon_{\overrightarrow{I^1_1}}(\text{Paul}_X, \epsilon_{\overrightarrow{I^1}}(\text{back}_E, (\text{will come})_{I^1}))), \epsilon_{\overrightarrow{I^1_1}}(\text{Liza}_X, (\text{will smile})_{I^1}))$$

(4-38b. — page 123) Liza will smile.

$$\epsilon_{\overrightarrow{I^1_1}}(\text{Liza}_X, (\text{will smile})_{I^1})$$

(4-38c. — page 125) When will Liza smile? When Paul will come back.

$$?_{\overrightarrow{I}}(\text{wh-}_{\overrightarrow{I>E}}(\epsilon_{\overleftarrow{I^1_1}}(\text{-en}_E, \epsilon_{\overrightarrow{I^1_1}}(\text{Paul}_X, \epsilon_{\overrightarrow{E>I^1}}(\text{back}_E, (\text{will smile})_{I^1}))), \Box_G),$$
$$\text{wh-}_{\overrightarrow{I>E}}(\epsilon_{\overleftarrow{I^1_1}}(\text{-en}_E, \epsilon_{\overleftarrow{I^1}}(\text{Liza}_X, (\text{will smile})_{I^1})), \Box_G))$$

(4-39a. — page 123) I wonder how many errors I did.

$$\text{h-}_{\overrightarrow{I^2_2}}(\epsilon_{\overrightarrow{I^2_2>O}}(\epsilon_{\overrightarrow{Y}}(\epsilon_{\overrightarrow{A}}(\text{-ow}_E, \text{many}_A), (\text{errors})_Y), \epsilon_{\overleftarrow{I^2_1}}(\text{I}_X, \text{did}_{I^2_{dep}})),$$
$$\epsilon_{\overrightarrow{I^2_1}}(\text{I}_X, \text{wonder}_{I^2_{gov}}))$$

(4-39b. — page 125) What time is it?

$$\text{wh-}_{\overrightarrow{I^2_2}}(\epsilon_{\overleftarrow{Y}}(\text{-at}_A, \text{time}_Y), \epsilon_{\overleftarrow{I^2_1}}(\text{it}_X, \text{is}_{I^2}))$$

(4-39c. — page 125) How many errors I did? I wonder.

$$?_{\overrightarrow{I}}(\epsilon_{\overrightarrow{I^1_1}}(\text{I}_X, \text{wonder}_{I^1}), \text{h-}_{\overrightarrow{I>A>O}}(\epsilon_{\overleftarrow{I^2_2}}(\epsilon_{\overrightarrow{Y}}(\epsilon_{\overrightarrow{A}}(\text{-ow}_E, \text{many}_A),$$
$$(\text{errors})_Y), \epsilon_{\overrightarrow{I^2_1}}(\text{I}_X, \text{did}_{I^2})), \Box_O))$$

(4-40a. — page 128) Karl giest die Blumen.

$$\epsilon_{\overleftarrow{I^2_2}}((\text{die Blumen})_Y, \epsilon_{\overrightarrow{I^2_1}}(\text{Karl}_X, \text{giest}_{I^2}))$$

(4-40b. — page 128) Ich lasse Karl die Blumen gießen.

$$\text{gießen}_{\overrightarrow{I^{2+1}_3}}((\text{die Blumen})_Y, \epsilon_{\overrightarrow{I^{2+1}_2}}(\text{Karl}_X, \epsilon_{\overleftarrow{I^{2+1}_1}}(\text{Ich}_Q, (\text{lasse})_{I^{2+1}_{cause}})))$$

(4-40c. — page 128) Ich lasse die Blumen gießen.

$$\text{gießen}_{\overrightarrow{I^{2+1-1}_3}}((\text{die Blumen})_Y, \epsilon_{\overrightarrow{I^{2+1}_2}}(\not{\text{Karl}}_X, \epsilon_{\overleftarrow{I^{2+1}_1}}(\text{Ich}_Q, (\text{lasse})_{I^{2+1}_{cause}})))$$

(4-41. — page 129) Paul makes Liza buy a car by Ron's for 8,000 pounds.

$$\text{for}_{\overleftarrow{I^{4+1}_5}}((\text{8,000 pounds})_W, \text{by}_{\overrightarrow{I^{4+1}_4}}((\text{Ron's})_Z, \text{buy}_{\overleftarrow{I^{4+1}_3}}((\text{a car})_Y,$$
$$\epsilon_{\overrightarrow{I^{4+1}_2}}(\text{Liza}_X, \epsilon_{\overrightarrow{I^{4+1}_1}}((\text{Paul})_Q, (\text{makes})_{I^{4+1}_{cause}})))))$$

(4-42a. — page 129) Him be a doctor!

$$\text{be}_{\overrightarrow{I^{2+1}_3}}((\text{a doctor})_Y, \epsilon_{\overrightarrow{I^{2+1}_2}}(\text{him}_X, \epsilon_{\overleftarrow{I^{2+1}_1}}(\not{\text{you}}_Q, \not{\text{let}}_{I^2_{cause}})))$$

(4-42b. — page 129) My mother ride a train!

$$\text{ride}_{\overleftarrow{\mathrm{I}_3^{2+1}}}((\text{a train})_{\mathrm{Y}}, \epsilon_{\overrightarrow{\mathrm{I}_2^{2+1}}}((\text{My mother})_{\mathrm{X}}, \epsilon_{\overleftarrow{\mathrm{I}_1^{2+1}}}(\cancel{\text{you}}_{\mathrm{Q}}, \cancel{\text{let}}_{\mathrm{I}^2_{\text{cause}}})))$$

(4-42c. — page 129) Her wash the dishes!

$$\text{wash}_{\overleftarrow{\mathrm{I}_3^{2+1}}}((\text{the dishes})_{\mathrm{Y}}, \epsilon_{\overrightarrow{\mathrm{I}_2^{2+1}}}(\text{her}_{\mathrm{X}}, \epsilon_{\overleftarrow{\mathrm{I}_1^{2+1}}}(\cancel{\text{you}}_{\mathrm{Q}}, \cancel{\text{let}}_{\mathrm{I}^2_{\text{cause}}})))$$

(4-43a. — page 131) (Let) her do it!

$$\text{do}_{\overleftarrow{\mathrm{I}_3^{2+1}}}(\text{it}_{\mathrm{Y}}, \epsilon_{\overrightarrow{\mathrm{I}_2^{2+1}}}(\text{her}_{\mathrm{X}}, \epsilon_{\overleftarrow{\mathrm{I}_1^{2+1}}}(\cancel{\text{you}}_{\mathrm{Q}}, \cancel{\text{let}}_{\mathrm{I}^2_{\text{cause}}})))$$

(4-43b. — page 131) (Help) him put his shoes!

$$\text{put}_{\overleftarrow{\mathrm{I}_3^{2+1}}}((\text{his shoes})_{\mathrm{Y}}, \epsilon_{\overrightarrow{\mathrm{I}_2^{2+1}}}(\text{him}_{\mathrm{X}}, \epsilon_{\overleftarrow{\mathrm{I}_1^{2+1}}}(\cancel{\text{you}}_{\mathrm{Q}}, \cancel{\text{help}}_{\mathrm{I}^2_{\text{cause}}})))$$

(4-43c. — page 131) (Make) this stay open!

$$\text{stay}_{\overleftarrow{\mathrm{I}_3^{2+1}}}((\text{open})_{\mathrm{Y}}, \epsilon_{\overrightarrow{\mathrm{I}_2^{2+1}}}((\text{this})_{\mathrm{X}}, \epsilon_{\overleftarrow{\mathrm{I}_1^{2+1}}}(\cancel{\text{you}}_{\mathrm{Q}}, \cancel{\text{make}}_{\mathrm{I}^2_{\text{cause}}})))$$

A.5 Constructive dictionaries

The examples in Chapter 5 are not reported since they are not 'reference' examples which can be extracted from the context they have been presented, and used to instance other cases on the same guidelines.

A.6 Discourse and beyond

(6-2a. — page 158) I've seen Mary with Paul. (*no emphasis*)

$$\epsilon_{\overrightarrow{\mathrm{I}_2^2}}(\text{with}_{\overleftrightarrow{\mathrm{A}}>\mathrm{Y}}(\text{Paul}_{\mathrm{Q}}, \text{Mary}_{\mathrm{Y}}), (\text{I've seen})_{\mathrm{I}_1^2})$$

(6-2b. — page 158) I've seen **Mary** with Paul. (*e.g., not Liza*)

$$\epsilon_{\overrightarrow{\mathrm{I}_2^2}}(\text{with}_{\overleftarrow{\mathrm{A}}>\mathrm{Y}}(\text{Paul}_{\mathrm{Q}}, \textbf{Mary}_{\mathrm{Y}}), (\text{I've seen})_{\mathrm{I}_1^2})$$

(6-2c. — page 158) I've seen Mary with **Paul**. (*e.g., not with John*)

$\epsilon^{\rightarrow}_{\mathrm{I}^2_2}(\text{with}^{\rightarrow}_{\mathrm{A}>\mathrm{Y}}(\textbf{Paul}_{\mathrm{O}}, \text{Mary}_{\mathrm{Y}}), (\text{I've seen})_{\mathrm{I}^2_1})$

(6-3a. — page 164) I won the race (*I am telling that I won the race*)

$\epsilon^{\leftarrow}_{\downarrow \mathcal{A}}(\epsilon^{\leftarrow}_{\mathrm{I}^2_2}((\text{the race})_{(\mathrm{Y},\mathrm{S})}, \epsilon^{\rightarrow}_{\mathrm{I}^2_1}(\text{I}_{\mathrm{S}}, \text{won}_{\downarrow \mathcal{A}})), (\text{assert})_{(\downarrow \mathcal{A},\mathrm{S})})$

(6-3b. — page 166) Win the race! (*I order you to win the race*)

$\epsilon^{\leftarrow}_{\uparrow \mathcal{D}}(\epsilon^{\leftarrow}_{\mathrm{I}^2_2}((\text{the race})_{(\mathrm{Z},\mathrm{R})}, \epsilon^{\leftarrow}_{\mathrm{I}^2_1}((\text{you})_{\mathrm{R}}, \text{win}_{\uparrow \mathcal{D}})), (\text{direct})_{(\uparrow \mathcal{D},\mathrm{S})})$

(6-3c. — page 168) I'll win the race (*I promise myself to win the race*)

$\epsilon^{\leftarrow}_{\uparrow \mathcal{C}}(\epsilon^{\leftarrow}_{\mathrm{I}^2_2}((\text{the race})_{(\mathrm{Z},\mathrm{S})}, \epsilon^{\leftarrow}_{\mathrm{I}^2_1}(\text{I}_{\mathrm{S}}, (\text{'ll win})_{\uparrow \mathcal{C}})), (\text{commit})_{(\uparrow \mathcal{C},\mathrm{S})})$

(6-3d. — page 170) I thank my team for their help.

$\epsilon^{\leftarrow}_{= \mathcal{E}}(\text{for}^{\leftarrow}_{\mathrm{E}}((\text{their help})_{(\mathrm{Y},\mathrm{S})}, \epsilon^{\leftarrow}_{\mathrm{I}^2_2}((\text{my team})_{\mathrm{R}}, (\text{I thank})_{(\mathrm{Y},\mathrm{S})})),$
$(\text{express})_{(= \mathcal{E},\mathrm{S})})$

(6-3e. — page 172) ...and the winner of the race is...

$\text{and}^{\leftarrow}_{\lambda}((\text{the winner of the race is...})_{\updownarrow \mathbb{D}}, (\mathit{Declare})_{\updownarrow \mathbb{D}})$

(6-4. — page 164) Speaker asserts that speaker's belief exists.

$\text{that}^{\leftarrow}_{\downarrow \mathcal{A}}(\epsilon^{\rightarrow}_{\mathrm{Y}}(\text{'s}^{\rightarrow}_{\mathrm{Y}}((\text{speaker})_{\mathrm{X}=\mathrm{S}>\mathrm{A}}, (\text{belief})_{\mathrm{Y}}), (\text{exists})_{\mathrm{I}^{\mathrm{x}}}),$
$\square^{\rightarrow}_{\downarrow \mathcal{A}}((\text{speaker})_{\mathrm{X}=\mathrm{S}}, (\text{asserts})_{\downarrow \mathcal{A}}))$

(6-5. — page 165) I tell you what I found in the library.

$\cdot^{\leftarrow}_{\downarrow \mathcal{A}}((\text{you what I found in the library})_{\uparrow \mathcal{D}}, (\mathit{I\_tell})_{\downarrow \mathcal{A}})$

(6-6. — page 166) Speaker directs receiver to do request.

$\epsilon^{\leftarrow}_{\uparrow \mathcal{D}}(\epsilon^{\rightarrow}_{\mathrm{I}^{\mathrm{x}}}((\text{request})_{\mathrm{Z}}, \epsilon_{\mathrm{I}^{\mathrm{x}}}((\text{receiver})_{\mathrm{Y}=\mathrm{R}}, (\text{to\_do})_{\mathrm{I}^{\mathrm{x}}})),$
$\square^{\rightarrow}_{\uparrow \mathcal{D}}((\text{speaker})_{\mathrm{X}=\mathrm{S}}, (\text{directs})_{\uparrow \mathcal{D}}))$

(6-7. — page 166) Liza ordered Paul to leave.

$$\epsilon^{\leftarrow}_{\uparrow \mathcal{D}}(\text{(Paul to leave)}_{(Z,R)}, \text{(Liza ordered)}_{(\uparrow \mathcal{D}, S)})$$

(6-8. — page 168) Speaker promises that speaker will do the promise.

$$\epsilon^{\leftarrow}_{\uparrow \mathcal{C}}(\epsilon^{\rightarrow}_{I^x}(\text{(promise)}_Z, \epsilon^{\leftrightarrow}_{I^x}(\text{(speaker)}_{Y=S}, \text{(will\_do)}_{I^x})), \Box^{\rightarrow}_{\uparrow \mathcal{C}}(\text{(speaker)}_{X=S}, \text{(promises)}_{\uparrow \mathcal{C}}))$$

(6-9. — page 169) Speaker expresses speaker's emotion on receiver.

$$\text{on}^{\leftarrow}_{I^x}(\text{(receiver)}_R, \text{'s}^{\rightarrow}_{I^x}(\text{(speaker)}_S, \text{(emotion)}_{I^x = Y})) \; \epsilon^{\rightarrow}_{= \mathcal{E}}(\text{(speaker)}_{X=S}, \text{(expresses)}_{= \mathcal{E}})$$

(6-10. — page 173) Speaker Declares: 'the Receiver has the Declared Status'.

$$:^{\leftarrow}_{\updownarrow \mathbb{D}}(\epsilon^{\leftarrow}_{I^x}(\text{status}_{\mathbb{S}}, \epsilon^{\rightarrow}_{I^x}(\text{receiver}_{Y=R}, \text{has}_{I^x})), \epsilon^{\rightarrow}_{\updownarrow \mathbb{D}}(\text{speaker}_{X=S}, \text{Declares}_{\updownarrow \mathbb{D}}))$$

(6-11. — page 174) (*Father:*) You (*son*) are the black sheep of the family.

$$\epsilon^{\leftarrow}_{\updownarrow \mathbb{D}}(\epsilon^{\rightarrow}_{I^2_2}(\text{of}^{\rightarrow}_{\mathbb{S}}(\text{(the family)}_O, \epsilon^{\leftarrow}_{O}(\text{the}_A, \epsilon^{\leftarrow}_{O}(\text{black}_A, \text{sheep}_O))), \epsilon^{\leftarrow}_{I^2_1}(\text{You}_R, \text{are}_{I^2})), \epsilon^{\rightarrow}_{\updownarrow \mathbb{D}}(\text{father}_S, \text{Declares}_{\updownarrow \mathbb{D}}))$$

(6-12. — page 175) Did you see the news?

$$?^{\leftarrow}_{\downarrow \mathcal{A}}(\text{(Did you see the news)}_{\downarrow \mathcal{A}}, \text{(assert)}_{\downarrow \mathcal{A}})$$

(6-13. — page 175) Why have you still to finish your homework?

$$?^{\leftarrow}_{\uparrow \mathcal{D}}(\text{(Why have you still to finish your homework)}_{\uparrow \mathcal{D}}, \text{(direct)}_{\uparrow \mathcal{D}})$$

(6-14. — page 175) On what I have my mind fixed today?

$$?^{\leftarrow}_{\uparrow \mathcal{C}}(\text{(Why have you still to finish your homework)}_{\uparrow \mathcal{C}}, \text{(commit)}_{\uparrow \mathcal{C}})$$

(6-15. — page 175) How should I feel about that?

$$?^{\leftarrow}_{= \mathcal{E}}(\text{(How should I feel about that)}_{= \mathcal{E}}, \text{(express)}_{= \mathcal{E}})$$

(6-16. — page 175) I promise you'll do it.

$\epsilon\overrightarrow{\ }_{\uparrow \mathcal{D}}((\text{you'll do it})_{(\uparrow \mathcal{D}, \mathrm{S})}, (\text{I promise})_{(\uparrow \mathcal{C}, \mathrm{S})})$

(6-17. — page 175) I request myself to be brave.

$\epsilon\overrightarrow{\ }_{\uparrow \mathcal{C}}((\text{to be brave})_{(\uparrow \mathcal{C}, \mathrm{S})}, (\text{I request})_{(\uparrow \mathcal{D}, \mathrm{S})})$

(6-18. — page 177) Why should I be angry with him?

$?\overleftarrow{\ }_{\uparrow \mathcal{D}}((\text{should I be angry with him})_{(= \mathcal{E}, \mathrm{S})}, (\textit{tell\_me}\ \text{why})_{(\uparrow \mathcal{D}, \mathrm{S})})$

(6-19. — page 177) I wonder if I am ashamed to follow her.

$\text{if}\overleftarrow{\ }_{\downarrow \mathcal{A}}((\text{I am ashamed to follow her})_{(= \mathcal{E}, \mathrm{S})}, (\text{I wonder})_{(\downarrow \mathcal{A}, \mathrm{S})})$

Appendix B

The Formal Model

The purpose of this appendix is to describe the formal model as a standalone mathematical entity.

The motivations behind its structure lie in linguistics and they have been described at wide and in depth in the main chapters of the present book. Here, we want to abstract over the linguistic context in order to show the formal model, which guided our investigations and enabled our findings, in its purest form.

The purpose of the formal model is to provide a natural framework to describe the basic structures of the grammars and the languages, focusing on the adpositional paradigm. The formal model has not been conceived to provide a parser for the language or an algorithm to generate all the well-formed productions of a language, although, to some extent, it can be used to obtain these results as side-effects.

In our experience, accumulated in the years devoted to study the adpositional approach and in writing this book, the formal model proved to be a formidable tool to clear thinking, as it often showed naturally and easily the correct way to interpret the most intricate and obscure linguistic phenomena.

Also, many unification results we have introduced in the text has been derived by firstly representing a phenomenon in the model and then, by observing that the obtained representation was wider and deeper than initially planned, eventually modelling other phenomena. So, we decided to include a complete mathematical presentation of the formal model, to provide the reader with the same instruments we used in pursuing our discoveries.

But we have to warn the reader that, in the following, deviating from the practice of previous chapters, we assume confidence with the fundamental concepts, definitions and results of basic Category Theory, and some degree of mathematical maturity.

Specifically, we assume as given the notions and results in Mac Lane (1998) and Borceux (1994), and we suggest some acquaintance with the notion of *sheaf* and *Grothendieck topology* (Johnstone, 2002; Mac Lane and Moerdijk, 1992) in order to track the inspiration and the possible future developments of the model.

B.1 Grammar categories

In the following we deal with general categories. As an intuitive guideline, the reader may think to objects in a category as adtrees and to arrows as constructions. This view, although inspiring, needs some work to be consistent, so we will introduce a series of definitions to fix the fundamental properties leading to the intended interpretation.

The fundamental property of grammar characters is that every object has exactly one grammar character. Also, we require that the set of grammar characters is meaningful, i.e., every grammar character contains at least one object.

Definition B.1.1 (Grammar character) *Given a small category $\mathbb{C}$, we say that G is a set of* grammar characters on $\mathbb{C}$ *if G is a partition on* $\operatorname{Obj}\mathbb{C}$*, the set of objects in $\mathbb{C}$, such that no class is empty.*

Formally, $G = \{G_i\}_{i\in GC}$ such that $\bigcup_{i\in GC} G_i = \operatorname{Obj}\mathbb{C}$ and, for any $i, j \in GC$, if $i \neq j$ then $G_i \cap G_j = \emptyset$, and $G_i \neq \emptyset$ otherwise.

In order to exactly define what we intend for constructions, we need the technical concept of grammar product.

Definition B.1.2 (Grammar product) *Fixed a small category $\mathbb{C}$, a set $G = \{G_i\}_{i\in GC}$ of grammar characters with $W \in G$, we call* grammar product on $\mathbb{C}$, G, W *any pair $\langle Y_1 \times \cdots \times Y_n, (z_1, \ldots, z_n)\rangle$ where n is a natural number called the* degree *of the product, $\times$ is the Cartesian product of sets and, for every i in $1 \ldots n$, $Y_i \in G \setminus \{W\}$, and $z_i \in W$.*

The reader is invited to notice that grammar products are not products in the category $\mathbb{C}$—we do not even require that $\mathbb{C}$ has products. Instead, they should be thought of as finite products of grammar characters, where the i-th projector is named by an element of W, which denotes the class of adpositions.

Constructions are families of arrows with the same domain and whose codomain lies in a single grammar character which is not the one of adpositions. These arrows are indexed by a grammar product.

Definition B.1.3 (Grammar construction) *Given a small category $\mathbb{C}$, a set $G = \{G_i\}_{i\in GC}$ of grammar characters with $W \in G$, a* grammar construction $\eta_{P,x}$ over P from *x is an indexed collection $\left\{f_j\right\}_{j\in P'}$ where*

- *$P = \langle P', a \rangle$ is a grammar product on $\mathbb{C}$, G, W;*
- *x is an object of $\mathbb{C}$;*
- *there is $V \in G \setminus \{W\}$ such that, for every $j \in P'$, $f_j \in \mathrm{Hom}_{\mathbb{C}}(x, y)$ where $y \in V$.*

We denote the set V by $\operatorname{cod} \eta_{P,x}$, *the codomain of the grammar construction. Similarly,* $\operatorname{dom} \eta_{P,x} = x$ *and we call P the product of $\eta_{P,x}$. Sometimes, we will write $\eta_{j,x}$ for f_j.*

A family $\{\eta_{P,x}\}_{x \in U}$ of grammar constructions over P from x varying in the U grammar character forms a grammar construction $\eta_{P,U}$ over P from U *when all its elements share the same codomain. We denote the common codomain with* $\operatorname{cod} \eta_{P,U}$. *Also,* $\operatorname{dom} \eta_{P,U} = U$ *and P is the product of the construction. Again, we will write $\eta_{j,x}$ to indicate the j-th component of the x-th component of $\eta_{P,U}$.*

We say that the construction η over $P = \langle Y_1 \times \cdots \times Y_n, (z_1, \ldots, z_n) \rangle$ on the object x, has x as *governor* and its *dependents* are the elements of $Y_1, \ldots, Y_n$; its adpositions are $z_1, \ldots, z_n$. So, its product P describes the adpositional structure that η imposes over the governor.

For symmetry, given a construction η as above, it is useful to consider the constructions which differ from η by exchanging a dependent with the governor.

Definition B.1.4 (Conjugate construction) *Fixed a small category $\mathbb{C}$, a set $G = \{G_i\}_{i \in GC}$ of grammar characters with $W \in G$, and a grammar construction*

$$\eta = \left\{ \left\{ f_j \colon x \to c(x, j) \right\}_{j \in P'} \right\}_{x \in V}$$

over the product $P = \langle P' = P_1 \times \cdots \times P_n, (a_1, \ldots, a_n) \rangle$ from the V grammar character, a grammar construction θ over the product Q from the grammar character U is said to be conjugate to η *if, for some k in $1 \ldots n$, $U = P_k$ and*

$$Q = \langle P_1 \times \cdots \times P_{k-1} \times V \times P_{k+1} \times \cdots P_n, (a_k, a_2 \ldots, a_{k-1}, a_1, a_{k+1} \ldots, a_n) \rangle$$

and, for all $(x, j_1, \ldots, j_n) \in C \times P'$, $\operatorname{cod} \eta_{(j_1,\ldots,j_n),x} = \operatorname{cod} \theta_{(j_1,\ldots,j_{k-1},x,j_{k+1},\ldots,j_n),j_k}$.

We notice that, in general, a construction may have any number of conjugates for each elements in its product, even zero.

Constructions from grammar categories can be composed, to obtain more complex constructions.

Definition B.1.5 (Composition of constructions) *Given a small category* $\mathbb{C}$*, a set* $G = \{G_i\}_{i \in GC}$ *of grammar characters with* $W \in G$*, and grammar constructions*

$$\eta = \left\{ \left\{ f_{(y_1,\ldots,y_n)} \colon x \to c(x, y_1, \ldots, y_n) \right\}_{y_1 \in P_1,\ldots,y_n \in P_n} \right\}_{x \in V}$$

over the product $P = \langle P_1 \times \cdots \times P_n, (a_1, \ldots, a_n) \rangle$ *from the* V *grammar character and*

$$\theta = \left\{ \{ g_{(z_1,\ldots z_m)} \colon x \to d(x, z_1, \ldots, z_m) \}_{z_1 \in Q_1,\ldots,z_m \in Q_m} \right\}_{x \in U}$$

over the product $Q = \langle Q_1 \times \cdots \times Q_m, (b_1, \ldots, b_m) \rangle$ *from the* U *grammar character, the* composition $\theta \circ \eta$ *is defined when* $\operatorname{dom} \theta = U = \operatorname{cod} \eta$*. In this case,*

$$\theta \circ \eta = \left\{ \left\{ g_{(z_1,\ldots z_m)} \circ f_{(y_1,\ldots y_n)} \right\}_{y_1 \in P_1,\ldots,y_n \in P_n, z_1 \in Q_1,\ldots,z_m \in Q_m} \right\}_{x \in V}$$

and $\operatorname{dom} \theta \circ \eta = \operatorname{dom} \eta = V$*,* $\operatorname{cod} \theta \circ \eta = \operatorname{cod} \theta$*, and the product of* $\theta \circ \eta$ *is*

$$\langle P_1 \times \cdots P_n \times Q_1 \times \cdots \times Q_m, (a_1, \ldots, a_n, b_1, \ldots, b_m) \rangle \quad .$$

As the reader may check, the grammar characters along with all the possible constructions and with the composition just defined, forms a category.

Although this category is of interest, as we will briefly discuss in the following, the kernel of the formal model concentrates upon another category.

Definition B.1.6 (Grammar category) *Consider a small category* $\mathbb{C}$*, a set of grammar characters on it* $G = \{G_i\}_{i \in GC}$ *with* $W \in G$*, and a set* C *of grammar constructions over some product on* $\mathbb{C}$*,* G*,* W *from a grammar character in* G*. Then, the structure* $\langle \mathbb{C}, G, W, C \rangle$ *is said to be a* grammar category *when it satisfies the following properties:*

1. *There is a collection* M *of objects in* $\mathbb{C}$*, called the* atoms of the category, *such that, for each* $x \in \operatorname{Obj} \mathbb{C}$*, there is an arrow* f *in* $\mathbb{C}$ *with* $\operatorname{dom} f \in M$ *and* $\operatorname{cod} f = x$*; moreover, if* $x \in M$*, then* $f = \mathrm{id}_x$*;*

2. *For each arrow* $f \colon a \to b$ *in* $\mathbb{C}$*, there is* $\eta_{P,V} \in C$ *such that* $a \in V$ *and* $f \in \eta_{P,a}$*, where* $\eta_{P,a}$ *is the* a*-component of* $\eta_{P,V}$*;*

3. *Every construction* $\eta \in C$ *is the composition of a sequence* $\theta_1, \ldots, \theta_n$ *of constructions in* C *each one having degree* 1*;*

4. *Each construction in C has exactly one conjugate construction for every element in its product.*

Informally, a grammar category is a category whose arrows can be grouped into a collection of constructions satisfying a number of regularity properties: the collection comprises all the arrows in the category; every object can be reached from a set of atoms; each construction has all its conjugates; every construction can be reduced to a sequence of one-step constructions.

In the whole book we dealt with a subclass of grammar categories.

Definition B.1.7 (Adpositional category) *A grammar category* $\mathbb{A}$ *is* adpositional *when* $\mathbb{A} = \langle \mathbb{C}, G, U, C \rangle$ *and* G *is indexed by*

$$GC = \{A, E, O, U\} \cup \left\{ I_p \colon p \text{ is a grammar product on } \mathbb{C}, G, U \right\} .$$

We require that the collection of I_p*s is wide enough to allow the definition of all the related constructions, mainly valency.*

B.2 Adtrees

Let $\langle \mathbb{C}, G = \{G_i\}_{i \in GC}, W, C \rangle$ be a grammar category and let η be a grammar construction over $\langle Y_1 \times \cdots \times Y_n, (a_1, \ldots, a_n) \rangle$ from the object x:

$$\eta = \left\{ f_{(y_1,\ldots,y_n)} \colon x \to c_{(x,y_1,\ldots,y_n)} \right\}_{y_1 \in Y_1,\ldots,y_n \in Y_n} .$$

We can equivalently write each arrow in η as[a]

$$x \xrightarrow{f_{(y_1,\ldots,y_n)}} a_n(y_n, a_{n-1}(\ldots, a_1(y_1, x)\ldots)) = c(x, y_1, \ldots, y_n) \in \operatorname{cod} \eta .$$

The above representation is evidently unique, as far as x is the governor and $(y_1, \ldots, y_n)$ are the dependents taken in the given order, so we have shown that every object in a grammar category can be represented as an adtree whose leaves are atoms. Of course, if we take a conjugate construction [1] η^*, the governor would be y_k and the dependents would be $(y_1, \ldots, y_{k-1}, x, y_{k+1}, \ldots, y_n)$, corresponding to the representation

$$a_n(y_n, a_{n-1}(\ldots, a_{k_1}(y_{k-1}, a_1(x, a_{k+1}(\ldots(a_k(y_1, y_k))\ldots))\ldots))$$

[a] We use linearised trees for compactness.

of the same object $c(x, y_1, \ldots, y_n)$.

To some extent, the converse also holds: if we take as objects the adtrees, as arrows the constructions described in the book along with their conjugates, we obtain an adpositional category. In this case, we have that a linguistic construction η generates a set of grammar constructions, itself plus the constructions which are conjugates to η.

It is interesting to notice that grammar categories, when thought of as categories of adtrees, give raise to a category of grammar characters.

Theorem B.2.1 *Fixed a grammar category $\langle \mathbb{C}, G = \{G_i\}_{i \in GC}, W, C \rangle$, if*

$$\eta = \left\{ \left\{ f_{(x,y_1,\ldots,y_n)} : x \to c_{(x,y_1,\ldots,y_n)} \right\}_{y_1 \in Y_1,\ldots,y_n \in Y_n} \right\}_{x \in V}$$

is a construction over $\langle Y_1 \times \cdots \times Y_n, (a_1, \ldots, a_n) \rangle$ from the grammar character V, then, for every $(y_1, \ldots, y_n) \in Y_1 \times \cdots \times Y_n$, consider

$$\theta_{(y_1,\ldots,y_n)} = \left\{ f_{(x,y_1,\ldots,y_n)} \right\}_{x \in V} \quad .$$

The structure having GC as objects and the θ's above as arrows, is a category

Proof: Immediate from the definitions, noticing that each θ is a set-theoretic function. ■

The meaning of the theorem is that we are allowed to construct new grammar categories literally over the old ones, by putting new grammar characters and new constructions over the category of the theorem. This operation is exactly what we have done when moving from the morphemic constructions to the adtree constructions in Chapter 4.

A property of grammar category which is important for computational purposes is that every object can be generated from atoms and constructions.

If a grammar category is such that every object can be represented by an adtree which is unique modulo conjugates, we will say that the category is *pure*. In this case, having a dictionary, which describes the set of atoms and the possible constructions, allows to generate all the possible adtrees.

In the adpositional case, it is evident that some adtrees will not correspond to well-formed linguistic productions, because, e.g., they will not agree on the gender. In general, all the adtrees generated via constructions from atoms will have a correct structure, but their leaves may not conform to the redundancies rules that every natural language has. In principle, it should be possible to cope with redundancies augmenting the number of grammar characters, as it is usual in the formal Chomsky grammars. But, in practice, this approach would multiply the number of constructions as well, obscuring the structural unity of a language. In this respect, we prefer to accept the generation of adtrees which fail to meet the redundancies of the language; we claim that it is possible to check redundancies in an adtree and to correct them after the generation of the adtree.

It is impossible to prove the previous claim except by an exhaustive analysis of the redundancies in the existing natural languages. And, we are not going to do that. Instead, we support the claim by observing that redundancy rules are always "local", that is, they apply on specific (and small, in practice) subtrees of a given adtree. Also, such rules can be expressed as a tree-pattern which induces a substitution on the match.

For example, a rule for articles *vs* nouns in Italian is *if the noun is feminine, the article must be feminine as well; if not, change it to a feminine article!*. We can formalise this rule as *if* $\epsilon(L, N)$ *is your adtree, and it is constructed via the "determiners" construction (see Chapter 3), and N has the "feminine" attribute, which is an attribute in the grammar category of morphemes, and L has not the "feminine" attribute, then substitute "la" for L.* Of course, in a complex adtree, we should apply recursively this rule to every subtree.

In this respect, the computational analysis of algorithms on adtrees can be somewhat simplified by the following result. We say that an adtree is *linear* when its governor and all its dependents are atoms. Moreover, if the adtree $a(x, T_0)$ is obtained by means of the η construction, then $a(T_0, x)$ is obtained by means of the η^* conjugate construction. We say that $T_1 \sim T_2$ when $T_1 = T[a(x, T_0)]$ and $T_2 = T[a(T_0, x)]$, i.e., when T_1 and T_2 differ by a single subtree which appears in the regular form in T_1 and in the conjugate form in T_2, and, evidently it holds that the object denoted by T_1 equals the object denoted by T_2. We denote by $\approx$ the reflexive, symmetric and transitive closure of the $\sim$ relation. Evidently, $\approx$ is an equivalence relation over adtrees. We call it equivalence modulo conjugates, or c-equivalence for short.

Theorem B.2.2 *Let $\mathbb{A}$ be a grammar category. Then, every object of $\mathbb{A}$ is denoted by an adtree x, and there is an adtree y, denoting the same object, such that $x \approx y$ and y is linear.*

Proof: We prove that every adtree T is c-equivalent to some linear adtree L denoting the same object as T. We proceed by induction on the structure of T:

- (Basis) If T is an atom, $L = T$ is linear and c-equivalent to T by reflexivity, and L and T denote the same object;

- (Induction step) If $T = a(T_2, T_1)$ then, by induction hypothesis, there are

$$L_1 = b_k(t_k, b_{k-1}(\dots, b_1(t_1, t_0)\dots))$$

and L_2, both linear, such that $T_1 \approx L_1$, $T_2 \approx L_2$, T_1 and L_1 denote the same object, and T_2 and L_2 denote the same object. Thus, $T \approx a(L_2, L_1)$ and they denote the same object, deriving from an application of the same construction to identical objects. So, considering the conjugate of the composed construction,

$$\begin{aligned} T &\approx a(L_2, b_k(t_k, b_{k-1}(\dots, b_1(t_1, t_0)\dots)) \approx \\ &\approx b_1(t_0, b_k(t_k, b_{k-1}(\dots, a(t_1, L_2)\dots)) = L \ , \end{aligned}$$

and L is linear and denotes the same object as T. ■

Corollary B.2.3 *In a grammar category, every adtree $T[T_1]$, i.e., an adtree having T_1 as a subtree, is c-equivalent to an adtree having T_1 as governor.*

Proof: The same induction argument proves the statement. ■

Corollary B.2.4 *Given a grammar category, the constructions can be partitioned into the basic ones and their conjugates. Then, every adtree constructed via basic constructions, is c-equivalent to a linear adtree and vice versa.*

Proof: A tedious analysis of the proof of the theorem, where we keep track of conjugates, allows to prove the result. ■

Corollary B.2.5 *In a pure grammar category, every object is uniquely identified with the quotient of the set of its adtrees via the $\approx$ relation.*

As a side effect of corollary B.2.4 we get an abstract parsing algorithm for any grammar category. In fact, a linear adtree is the same thing as a sequence of morphemes where adpositions, which can be mechanically extracted, are put in the "right" places. So, the meaning of corollary B.2.4 is that it is possible to reconstruct from such a linear adtree another adtree which is built by basic constructions only, i.e., an adtree in the format we used along all the book. Of course, such an algorithm is "universal", since it applies to any grammar category, but desperately inefficient, since it is highly non-deterministic.

Such a parsing algorithm is unable to distinguish different representations of the same objects unless they are c-equivalent. So, it should be used only on pure grammar categories. By the way, all the grammar categories we dealt with in this monograph are pure.

B.3 Transformations

Several times it has been said that indicators are essential to define and to understand transformations. The reader may have noticed that we have no space for indicators, or other attributes of adpositions and morphemes in the formal model. In fact, there is no need for a distinct treatment of linguistic features like indicators, as well as for the attributes (like gender or number) regulating the redundancies of the language.

Let us consider the case of indicators: if U is the set of adpositional morphemes in the language, it suffices to take $\overline{U} = U \times \{\leftarrow, \rightarrow, \leftrightarrow\}$ as the set of adpositions in the grammar category to ensure that each adposition is decorated by an indicator. Other attributes are treated similarly.

Moreover, we usually require that there is a special morpheme, ϵ, in the set of adpositions, representing the empty adposition. We also require that $\square$ is an element of each grammar character, representing the empty element of that class; to be precise, we should speak of $\square_A$, $\square_E$, $\square_O$, ...which are distinct objects in the grammar category, but the ambiguous $\square$ symbol does not cause any harm in the usual practice as it is always in a context clarifying its correct meaning.

Furthermore, we frequently used cancelled atoms to denote objects which are present in a linguistic structure but they are absent from its textual representation—we said they are "hidden". Again, these elements require no special treatment in the definition of grammar categories: if M is the set of atoms in the language, it suffices to take $\overline{M} = M \sqcup M$, the disjoint union of M with itself, as the set of atoms in the grammar category. The first copy of M stands for the atoms, while the second one is for the cancelled atoms; constructions and grammar characters are defined accordingly, making no distinction whether an atom is cancelled.

These manipulations on atoms and adpositions are silently assumed in the following, as they are frequently used to encode transformations. Also, they are transparent to the formal machinery that constitute the basis of the formal model.

Informally, a transformation is a map from adtrees to adtrees in a given grammar category.

As we have seen in section B.2, fixed a grammar category $\mathbb{C}$, an adtree is a representation of an instance of a construction which makes explicit its product, its dependents and its governor, organising them in a (binary) tree structure. Also, every object in $\mathbb{C}$, except for adpositions, can be represented by an adtree, and usually by more than one.

We will call $\mathrm{Ad}(\mathbb{C})$ the collection of all adtrees in $\mathbb{C}$, plus the empty adtree, denoted by $\perp$; in fact, $\mathrm{Ad}(\mathbb{C})$ is a category whose arrows are the arrows of $\mathbb{C}$, i.e., the instances of the constructions of $\mathbb{C}$, thought of as applied to adtrees instead of being applied to objects. Moreover, we assume that there is a unique arrow from $\perp$ to any other adtree on $\mathbb{C}$, so $\perp$ is an initial object in $\mathrm{Ad}(\mathbb{C})$.

Definition B.3.1 (Transformation) *Given a grammar category* $\mathbb{C}$*, a* transformation on $\mathrm{Ad}(\mathbb{C})$ *is an endofunctor in* $\mathrm{Ad}(\mathbb{C})$.

The previously described redundancy transformations satisfy the above definition. In general, all the "transformations" we considered in this book are transformations in the defined formal sense, except for "ϵ-transformations", which, as it has been remarked, are just a convenient way to condense adtrees, that is, a matter of human readability.

It is worth describing in some detail how redundancy transformations operate, both to clarify how they should be conceived and understood, and to introduce in a formal way the notion of correct adtrees with respect to redundancies.

As an example, consider the redundancy rule which says "if the noun is feminine, modify the article to feminine". In the grammar category for the Italian language, this rule becomes a functor F on adtrees as follows: an adtree T becomes the $F(T)$ adtree where every occurrence of a subtree $\epsilon(a, n)$ with a article and n feminine noun is substituted with $\epsilon(a', n)$ with a' the feminine article; every arrow, i.e., every instance of some construction, is mapped as obvious.

Since every redundancy transformation is analogous to the shown case, the example is completely general. In fact, every such transformation can be reduced to the substitution of an adtree pattern with a variant of its instance. Functoriality ensures the correct spreading of the reduced definition in complex adtrees.

Given an endofunctor F, we say that an adtree T is a *fixed point of* F if $T = F\,T$. Also, we say that the f adtree is *final for* F *on the adtree* t if f is a fixed point of F and $f = F \circ \cdots \circ F\,t$, i.e., if f is the result of a finite composition of F with itself applied to t. The meaning of this definition is that a transformation F is applied to an adtree t to "correct" its "wrong" redundancies; the transformation is iteratively applied until all the "mistakes" are removed and the resulting adtree does not change anymore.

The set of final adtrees for a redundancy transformation R deprived of $\perp$, the empty adtree, if present, is the set of adtrees which are *correct with respect to* R. Usually, we associate to every grammar category a set T_R of transformations which are responsible for discriminating the *correct adtrees for the grammar category*, which is the set of adtrees correct with respect to each functor in T_R.

In the book appears another type of transformations, the ones used to model complex constructions like auxiliary verbs or the relation between active and passive tenses. These transformations are endofunctors on the category of adtrees of a peculiar class of grammar categories. In some cases, it suffices to consider an adpositional category describing the language, like the English grammar category and the active-passive transformation.

In other cases, the linguistic phenomena under examination can be explained only considering the interplay between the linguistic level, modelled by the linguistic adpositional category, and the corresponding morphemic level, modelled by the morphemic adpositional category.

It has been claimed and explained that phenomena like the participle is a sort of morphemic summary for a more complex linguistic construction which is semantically equivalent. In the following, we will define the exact mathematical framework that enables a formal and precise definition of these phenomena as formal transformations.

As explained in Chapters 2, 3 and 4, the linguistic and morphemic levels of a natural language are modelled by two distinct adpositional categories. The atoms of the linguistic category are the objects corresponding to the correct adtrees for the morphemic category, in the formal sense previously explained. In fact, exploiting this link into an operation that combines the two categories into one, is the mathematical environment allowing to represent the phenomena of interest as transformations, i.e., endofunctors on the combined category.

Assume to have a grammar category $\mathcal{M} = \langle \mathbb{M}, G_M, W_M, C_M \rangle$, called the *category of morphemes*, and another grammar category $\mathcal{L} = \langle \mathbb{L}, G_L, W_L, C_L \rangle$, called the *linguistic category*, and a map $\phi\colon \operatorname{Obj}\mathbb{M} \to \operatorname{Obj}\mathbb{L}$ whose image lies in the atoms of $\mathcal{L}$. Define $C = \langle \mathbb{C}, G, W, C \rangle$ where

- the category $\mathbb{C}$ has as objects $\square_M$ plus $\operatorname{Obj}\mathbb{L} \sqcup \operatorname{Obj}\mathbb{M}$, where $M = \{\square_M\}$

and $\Box_M$ is not an object of $\mathbb{L}$ and $\mathbb{M}$; the arrows of $\mathbb{C}$ are the arrows of $\mathbb{L}$ and those of $\mathbb{M}$ plus the instances of ϕ;

- the atoms of C are, clearly, the atoms of $\mathcal{M}$ plus $\Box_M$;
- the grammar characters of $\mathcal{M}$ are $G = G_L \sqcup G_M \sqcup M$;
- the adpositions of C are $W = W_L \sqcup W_M \sqcup \{\epsilon_M\}$;
- the constructions of C are

$$C = C_L \sqcup C_M \sqcup \left\{ \left\{ \left\{ \theta_j \colon x \to \phi(x) \right\}_{j \in M} \right\}_{x \in \alpha} : \alpha \in G_M \right\} .$$

Evidently, C is a grammar category.

Informally, C corresponds to the disjoint union of $\mathcal{M}$ and $\mathcal{L}$ where the atoms of $\mathcal{L}$ are substituted, modulo an ϵ-transformation which hides the θ construction, by the morphemes, i.e., the objects of $\mathcal{M}$, via the ϕ correspondence.

In the C grammar category, every endofunctor of $\mathcal{L}$ and $\mathcal{M}$ can be trivially extended to act on C, so to become a transformation on C. Extending the redundancies transformations of $\mathcal{M}$ and $\mathcal{L}$, and considering their union, one naturally defines the concept of correct adtrees on C.

But, the additional endofunctors, the ones not being extension of endofunctors on $\mathcal{L}$ or $\mathcal{M}$, are precisely those that are used to provide the formal counterpart of the transformations we want to model, the ones described in Chapter 4.

B.4 An abstract view

In this section, we want to introduce the preliminaries of a possible future development of the formal model. The idea is to abstract over the concrete details in the definition of a grammar category, and to consider the overall structure from an external point of view.

The core of the abstract view is to consider the language as a formal space where constructions take place. Thus, studying a grammar means to understand the inner relations among constructions in a structural way. To achieve this goal, we follow the rather standard mathematical technique of looking at a category as a space via the so-called Grothendieck topologies. Then, we are allowed to study the inner structure of the obtained site via the sheaves on that site.

We will not develop here the whole mathematical machinery, which is rather complex, but we want to give the reader an intuition of what to expect and why we believe this approach could be of interest.

We recall that a *sieve on an object* c in a category $\mathbb{C}$ is a set S of arrows of $\mathbb{C}$ with codomain c such that, if $f \in S$ and $f \circ h$ is defined, then $f \circ h \in S$. Specialising this notion to pure grammar categories, we get that a sieve S on c is a set of instances of constructions in bijective correspondence with a set S' of adtrees defined as follows: if $f: b \to c \in S$, then $b \in S'$ and, for every d such that there is $g: d \to b$, $d \in S'$. Intuitively, a sieve S on c fixes a set B of instances of constructions of c, whose governors are in S', and the sieve contains all the instances of constructions of c whose last step is in B.

Dualising, a *cosieve* S *on* c in a pure grammar category is a set of instances of constructions of c in bijective correspondence with a set S' of adtrees defined as follows: if $f: c \to b \in S$, then $b \in S'$ and, for every adtree d whose governor is b, the adtree $d' \in S'$, where d' is obtained by d expanding the construction of b so to make c its governor. In other words, cosieves are built fixing a set B of constructions having c as governor, and taking all the constructions which are extensions of those in B.

Examples of sieves are as follows: suppose c is the result of a valency construction over the product $\langle O, (\epsilon) \rangle$ from v in some adpositional category; then, the sieve S generated by this arrow contains all the constructions of c having v as governor, eventually substituted by its constructions. Similarly, the cosieve on the object v generated by the arrow as above, is nothing else that the collection of instance of constructions having the arrow as the first step, i.e., it is isomorphic to the set of phrases whose principal sentence is the given verb construction.

Also, considering all the constructions obtained by adding a fixed circumstantial to an element of the I grammar characters, we obtain a sieve whose base arrows are the conjugates of the I-modifier construction.

In general, in a pure grammar category, fixing a collection of instances of constructions of the same object, a sieve contains all the (instance of constructions of the) objects represented by adtrees obtained by expanding the governors of those instances. On the other side, if we fix a collection of adtrees all having the same governor, the elements of the corresponding cosieve are all the (instance of constructions of the) adtrees whose governors are the fixed ones. In a general grammar categories, sieve and cosieve are understood in a similar way, except that the adtree representation is not unique modulo conjugates.

As sieves and cosieves represent interesting collections of (instances of) constructions, it is worth considering the meaning of Grothendieck topologies in the context of pure grammar categories. [2] We recall that a *Grothendieck topology on a category* $\mathbb{C}$ is a function J mapping every object c of $\mathbb{C}$ into a collection $J(c)$ of sieves on c such that

1. $t_c = \{f : \operatorname{cod} f = c\} \in J(c)$, i.e., the maximal sieve on c is always in $J(c)$;
2. if $S \in J(c)$ then, for every $h : d \to c$ in $\mathbb{C}$,
$$h^*(S) = \{g : \operatorname{cod} g = d \text{ and } h \circ g \in S\} \in J(d) \enspace ;$$
3. if $S \in J(c)$ and R is any sieve on c such that $h^*(R) \in J(d)$ for all $h : d \to c$ in S, then $R \in J(c)$.

A *site* is a pair $\langle \mathbb{C}, J \rangle$ where J is a Grothendieck topology on $\mathbb{C}$. Intuitively, a site is a topological space built on a category instead of a set. The reader should think to such a space as a potentially pointless space, where the basic elements are neighbours not necessarily containing points, which may be absent. Although these mathematical structures are not immediate to the geometrical intuition, their properties are essentially the same as the ones of usual spaces, only in a more abstract setting.

To clarify the meaning of a Grothendieck topology on a pure grammar category, one may think of a sieve $S \in J(c)$ as a set of admissible instances of constructions with respect to J. So, $h^*(S)$ becomes the set of instances of constructions that, composed with h, give an admissible construction of the object c. Hence, the stability axiom, i.e., the second requirement in the definition of a topology, is nothing more than requiring that $h^*(S)$ is admissible, which is a very natural condition. The third requirement amounts to a generalised transitivity. Thus, composing admissible constructions with themselves produces admissible constructions, which is the meaning of transitivity.

As a consequence, a Grothendieck topology J on a pure grammar category $\mathbb{C}$ is a way to coherently identify admissible instances of constructions. But the *sheaves on a site* $\langle \mathbb{C}, J \rangle$ form a topos $\mathrm{Sh}(\mathbb{C}, J)$ that contains the whole amount of information on the site. We remind that a *presheaf on* $\mathbb{C}$ is simply a functor $\mathbb{C}^{\mathrm{op}} \to \mathbf{Set}$ and a *sheaf on* $\langle \mathbb{C}, J \rangle$ is a presheaf P on $\mathbb{C}$ such that, for every object c in $\mathbb{C}$ and every sieve $S \in J(c)$, the inclusion $S \rightarrowtail \mathrm{Hom}(-, c)$ induces an isomorphism $\mathrm{Hom}(S, P) \cong \mathrm{Hom}(\mathrm{Hom}(-, c), P)$.

We do not want to explain here the details of what a sheaf is in a pure grammar category and the precise consequences of $\mathrm{Sh}(\mathbb{C}, J)$ being a topos, as these explanations are far beyond the scope of the present text. It suffices to say that, if we think of admissible constructions as a sort of abstract syntax for the language modelled by the category $\mathbb{C}$, then the topos $\mathrm{Sh}(\mathbb{C}, J)$ is able to provide a logical description for that syntax, which becomes an abstract grammar for the language, as well as a geometrical interpretation of the language as a mathematical space, which is a novel and different way to look at linguistic phenomena.

We believe that the analysis of $\mathrm{Sh}(\mathbb{C}, J)$ when $\mathbb{C}$ is a pure grammar category will be a difficult, but deeply rewarding future development for the formal model that may undercover and explain many subtleties of natural languages, and that may lead to a deeper understanding and unification of linguistic phenomena on a completely abstract basis, founded on the pure topos-theoretic framework.

B.5 Links with other formal models

In this section we want to clarify the relation between our formal model and context-free languages. Although the relations we draw can be generalised to context-sensitive languages and, beyond, to all decidable languages (and most of their sub-classes), we limit ourselves to the context-free case for clarity.

In the following, we assume acquaintance with the basic theory of formal languages. As a reference text, we will follow Martin (1997). Firstly, we recall the fundamental definition of context-free grammar and language. Then, we will show how to characterise a context-free language as a subset of objects in a suitable grammar category. The proposed representation is general and it can be easily adapted to wider and richer classes of languages and grammars. After, we will present a more specific construction which embodies the fundamental character of context-free languages, and provides a rather different characterisation of those languages in our formal model.

Definition B.5.1 *A* context-free grammar *is a structure* $\mathbb{G} = \langle V, \Sigma, S, P\rangle$ *where* V *and* Σ *are disjoint finite sets,* $S \in V$ *and* P *is a finite set of expressions of the form* $A \to \alpha$*, where* $A \in V$ *and* $\alpha \in (V \cup \Sigma)^*$*. The elements of* V *are called the* non-terminal symbols*, those of* Σ *are called the* terminals*.* S *is called the* start symbol *and the elements of* P *are called the* productions *or* derivation rules.

Definition B.5.2 *Let* $\mathbb{G} = \langle V, \Sigma, S, P\rangle$ *be a context-free grammar. The* language generated by $\mathbb{G}$ *is*

$$L(\mathbb{G}) = \left\{x \in \Sigma^* : S \xrightarrow{\mathbb{G}} x\right\} ,$$

where $\stackrel{\mathbb{G}}{\longrightarrow}$ *is the reflexive and transitive closure of the* $\blacktriangleright$ *relation which, given* $\alpha, \beta \in (V \cup \Sigma)^*$*, is defined as* $\alpha \blacktriangleright \beta$ *if there is* $v \in V$ *such that* $\alpha = \alpha' \cdot v \cdot \alpha''$ *and* $\beta = \alpha' \cdot \beta' \cdot \alpha''$ *where* $(v \to \beta') \in P$.

A language L is said to be context-free *if there is a context-free grammar* $\mathbb{G}$ *such that* $L = L(\mathbb{G})$.

Let $\mathbb{G} = \langle V, \Sigma, S, P\rangle$ be a context-free grammar. Consider the relation $\stackrel{\mathbb{G}}{\longrightarrow} \subseteq V \times (V \cup \Sigma)^*$; by definition, it is reflexive and transitive, so it naturally induces a composition operation. Define $\mathbb{C}'$ as the category whose objects are the strings σ on the $V \cup \Sigma$ alphabet such that $S \stackrel{\mathbb{G}}{\longrightarrow} \sigma$, and whose arrows are exactly the instances of the $\stackrel{\mathbb{G}}{\longrightarrow}$ relation.

Now, define $\mathbb{C}$ as the category $\mathbb{C}' \sqcup \mathbf{1}$ where $\sqcup$ is the coproduct of categories and $\mathbf{1}$ has only one object, ϵ, with its identity as the unique arrow.

Le G be the discrete partition of $\mathrm{Obj}\,\mathbb{C}$, whose classes are $\{c\}$ for every object c. Obviously, G is indexed by $\mathrm{Obj}\,\mathbb{C}$. Fixing as C all the instances of the $\stackrel{\mathbb{G}}{\longrightarrow}$ relation, one obtains in a trivial way that $\overline{\mathbb{G}} = \langle \mathbb{C}, G, \{\epsilon\}, C\rangle$ is a grammar category [3] whose atoms are $\{S, \epsilon\}$.

The language $L(\mathbb{G})$ coincides with the set of objects in $\overline{\mathbb{G}}$ which are not the domain of any arrow except identities, as it is immediate to prove. [4] The corresponding adtrees represent the possible constructions as sequences of applications of rules in P.

In a different, more abstract way, one could equivalently say that $L(\mathbb{G})$ is the collection of maximal arrows in the maximal cosieve on S. This characterisation in not limited to context-free languages, but applies to all families of languages in the Chomsky's hierarchy. [5]

Moreover, the reader is invited to notice how this characterisation relates to the discussion on "correct" adtrees and redundancy transformations: in fact, we can equivalently say that $L(\mathbb{G})$ is isomorphic to the set of non-empty adtrees in $\overline{\mathbb{G}}$ which are fixed points with respect the transformation mapping each adtree denoting $c \in \Sigma^*$ whose governor is S into itself, and each other adtree into the empty adtree.

Unambiguous context-free languages can be characterised in a different and more specific way as special elements in a grammar category. Firstly, suppose Λ, the empty string of terminal symbols, is not in $L(\mathbb{G})$, where $\mathbb{G} = \langle V, \Sigma, S, P\rangle$ is some unambiguous context-free grammar.

A standard result states that, if $\mathbb{G} = \langle V, \Sigma, S, P\rangle$ is a context-free grammar, then there is another context-free grammar $\mathbb{G}' = \langle V', \Sigma, S, P'\rangle$ such that $L(\mathbb{G}) = L(\mathbb{G}')$ and $\mathbb{G}'$ is in Chomsky normal form, i.e., every production in p' is either of the form $A \to B \cdot C$ with $B, C \in V'$, or of the form $A \to a$ with $a \in \Sigma$.

Define $\mathbb{C}^*$ as the category having as objects the strings on Σ^* and as arrows the derivation rules defined as follows, along with constructions and grammar characters:

- if $(A \to a) \in P'$ then $a \in G_A$;
- if $(A \to B \cdot D) \in P'$ and $b \in G_B$, $d \in G_D$, then $b \cdot d \in G_A$; moreover, the arrows $b \to b \cdot d$ and $d \to b \cdot d$ are in $\mathbb{C}^*$. Finally, these arrows are instances of the obvious constructions over the products $\langle G_D, (\epsilon)\rangle$ from G_B, and over $\langle G_B, (\epsilon)\rangle$ from G_D, respectively. These construction are in the set C of constructions. We observe that the latter construction is the conjugate of the former one.

Now, call $\mathbb{C}$ the full subcategory of $\mathbb{C}^*$ whose objects are in $\left(\bigcup_{v\in V'} G_v\right) \cup \{\epsilon\}$. Evidently,

$$\left\langle \mathbb{C}, \{\epsilon\} \cup \bigcup_{v\in V'} G_v, \{\epsilon\}, C \right\rangle$$

is a grammar category, and $L(\mathbb{G}) = L(\mathbb{G}') = G_S$. In a trivial way, the correct adtrees are defined as the non-empty fixed points of the transformation which maps every adtree denoting an object in G_S into itself, and any other adtree into the empty one.

It is interesting to notice that the adtrees in this grammar category, when only conjugate constructions are allowed, are exactly the usual derivation trees in the Chomsky tradition.

Notes

Chapter 1

[1] As the reader will see throughout the book, our morphosyntactic framework has important consequences in semantics and even more at a pragmatic level. Phonology is not considered here distinctively, except when it is a vehicle of pragmatic purposes.

[2] When talking about speakers, informants, and so on, we will use the masculine genre for the sake of simplicity, confident that this fact will not prejudice readers belonging to the feminine genre.

[3] In fact, this is not exact: Appendix B describes the mathematical structure of the formal model we adopt. We really mean that there is no strict division between the linguistic description and the formal counterpart in the main text, as such distinction would obscure the real content of this monograph.

[4] Some authors prefer the term 'syntacticization' rather than grammaticalization. We believe that syntacticization is not a good term as morphology is also involved—not to mention phonotactic phenomena.

[5] An analysis of similarities and differences of the pioneers of the Cognitive Revolution can be found in Allen Harris (2010) in Kibbee (2010), who quoted Thagard (2005) in Derry et al. (2005).

[6] Even Fillmore's Frame Semantics—which is the most developed grammar formalisation within the cognitive linguistics approach—largely neglects morphosyntactic facts. For an account, see at least Fillmore (1982) in Geeraerts et al. (2006).

[7] A notable exception is Langacker (1995) in Dondelinger et al. (1995), which enlightens how much Tèsniere's Structural Syntax is compatible within the cognitive grammar tenet.

[8] We are in debt with that work and the person that explained Tèsniere's ideas to us, that is, Fabrizio A. Pennacchietti, who based his syntactic analysis of prepositional systems on Ceccato (1961) and Brøndal (1940), besides Tesnière (1959) classic work. Pennacchietti has worked on prepositions in a unique and original way for more than three decades—and still continues. The most complete attempt to formalise his prepositional systems is present in Gobbo (2009); however, his taxonomy at last relies in his linguistic intuition and hence it cannot be fully formalised. Most of the works published by Pennacchietti are in languages other than English. An exception is Pennacchietti (2009) in Shamuel (2009), an application of his model to Classic Syriac and Sureth, where readers

can be introduced to the general model as well. An up-to-date version fully applied to Italian prepositions can be found in Pennacchietti (2006), while the early system—quite different, but by no means without interest—was applied to Esperanto back in Pennacchietti (1976). There are few actual applications of Pennacchietti's perspective on prepositions: Tosco (2006) in Borbone et al. (2006) applied it—with some adaptation—to Gawwada. For a presentation of the main results in terms of formalisation in English, Italian and Esperanto, see Gobbo (2010).

[9] Phonetics and phonology are closely related to the natural language they belong to, so their treatment is left as a further work.

[10] The unaccusative hypothesis was fully developed by Perlmutter and Postal (1984) in Perlmutter and Rosen (1984). While the distinction between unaccusative and unergative verbs was enlightening for us, the Universal Alignment Hypothesis was controversial since at least 1984—see, e.g., Goldberg (1995). The refinement of the Universal Alignment Hypothesis in a Chomskyan perspective, i.e., the Universal Theta Assignment Hypothesis (UTAH), was put in discussion even among Chomskyan linguists, who do not agree about its formulation and consequences—see at least Baker (1997) in (Haegeman, 1997, 105–121). On the other hand, the graphs proposed in the Relational Grammar seem to be unfit to be properly formalised. Moreover, the theory eventually brings to Arc Pair Grammar, which is "extremely heavy going largely because of its difficult-to-read relational networks and bizarre terminology" (Blake, 1990, 21).

[11] In this respect, it is worth mentioning the notion of *grammar product* as developed in Appendix B which is the formal counterpart of semantic roles in our formal model.

[12] See Croft (2009) for an extensive treatment of this particular case, and Chapter 5 for the notion of sememe.

[13] For a survey of the literature in Type Logic Grammar, see Morrill (2007).

[14] When a specific language is fixed, then our model provides a way to encode it in logical terms, as done for Esperanto in Gobbo (2009), which was chosen as a special case, being both a living and a planned language, showing high degree of regularity and a low degree of grammaticalization and hence of idiomatic expressions—which do exist indeed, as shown in the books about phraseology by Fiedler (1999, 2002).

[15] Technically, the adtree on the left of Figure 1.1 is obtained by applying a conjugate construction, see Appendix D. We chose this representation instead of the canonical construction to emphasise the symmetry between the two possible constructions.

[16] Correctness in the case of adpositional trees is slightly more complex to check: in fact, it is possible and convenient to allow the formation of incorrect adtrees as far as we can discriminate the correct ones in a formal way. We will treat this aspect using the so-called *redundancy transformations*.

[17] Unfortunately, the word category has a double technical meaning: in Mathematics it is an algebraic structure, whereas in computational linguistics it refers to a class of syntactically decorated expressions. In the first case, we will speak of mathematical categories, while we use the term linguistic categories for the second usage. We will omit the specification when it is clear from the context.

[18] We make a slight deviation from the standard practice of Category Theory where constructions are identified with arrows, and arrows have unique starting and ending points. Instead, we use the word 'constructions' for distinguished and uniform classes of arrows, following the natural usage in linguistics.

[19] In Appendix B we will sketch an even deeper and by far more interesting topos-theoretic interpretation of our formal model.

[20] In this context, the word 'category' is usually understood as linguistic category. But in most formalisms linguistic categories are nothing else that the object of a mathematical category. This follows from the fact that the models of λ-calculus are Cartesian closed categories, see Lambek and Scott (1986), and that most formalisms can be interpreted as instances of some typed λ-calculus.

[21] In fact, this is an instance of one of the most compelling open problems in Mathematics and Computer Science: the P *vs.* NP question, see Garey and Johnson (1979).

Chapter 2

[1] We will explain in Chapter 5 our treatment of words in terms of morphology. By the moment, we consider words as indivisible units for the reader's sake.

[2] Note that, unlike the common usage, we use minimal pairs at any linguistic level, not only phonology.

[3] The reader should notice that the scene does not change, although a simplified expression is used to describe it. So, although 'good sold here' may be a valid English expression—as 'good' is also a noun—it does not describe the same scene, so it is invalid.

[4] Here, we assume the simplified view that every (formal) construction is uniquely defined by its grammar product, see Appendix B for the details. In general, one should say that the dependent selects the specific arrow among the construction over some product from the governor.

[5] In the formal terms described in Chapter 1, one may think that there is a 'rule in the grammar' which sends the *gov* adtree into the adtree having *gov* as right-branch, *dep* as left branch and *adp* as root. Also, the complete adtree belongs to the grammatical class *gc*; grammatical classes are used to control the construction of adtrees.

[6] It is obvious why the triangles are useful: without them, a completely explicit adtree of a complex real-world case can easily fill all the paper, but—even more importantly—it does not help the reader to find the aspects we are dealing with in that particular moment.

[7] Using the terminology introduced by Langacker (1987), the information prominence is where the focus of attention is placed, and it is called *trajectory* (*tr*), while the other element is called the *landmark* (*lm*). If something emerges from the landmark, it will be a secondary trajectory, so that this asymmetry can be activated again recursively. It is worth noticing that within cognitive linguistics this asymmetry (*tr*/*lm*) is used to define the *semantics* of the construction, while within the adpositional paradigm this asymmetry is used to construct the adtrees, i.e., the *syntax*. To avoid name clashing, we choose to avoid the use of *tr* and *lm* hereby, preferring to use instead the expressions 'information prominence', 'prominence', or 'prominent'.

[8] For a discussion of how the joint attentional frame works at a cognitive level, see in particular Tomasello (2003).

[9] It is worth noticing that linguistic meaning cannot be reduced *tout court* to a fully compositional and computational model. Meaning involves not only the *construable*, i.e., the process whereby concepts are structured in a given construction, but it also involves the *content* evoked, which is encyclopædic in scope. For example, the different way of counting hotel levels in the US and in USSR pertains to the content, not to the construable. Constructive adpositional grammars deal with the construable, not the content. We will come back on this issue when we will treat semantics.

[10] At a first glance, it may seem strange to apply the concept of government to two different sentences, which are normally considered to be independent. This assumption derives from the fact that the Chomskyan approach has mainly focused on the *intra*-phrasal level, letting the textual phenomena at the periphery of modelling, and thus considering them as independent. We argue that this assumption can bring severe limits to the expressive power of the model upon which it is built. By contrast, in the adpositional paradigm constructions can occur at *any* major linguistic level: morphological (intra-phrasal), syntactic (inter-phrasal), and textual (between sentences).

[11] Dependency is a concept which has a long and complex history in linguistic modelling, starting in modern times from Tesnière (1959). The theoretical assumption behind all dependency-based models is that "the syntactic structure of sentences resides in binary asymmetrical relations holding between lexical elements." (Nivre, 2005, 6). Unfortunately, there is no agreement among scholars about the use of the notion of dependency. For example, in Topological Dependency Grammar (TDG), proposed by Debusmann (2001), there are two different forms of dependencies, called 'syntactic dependency tree (ID tree)' and 'topological dependency tree (LP tree)'. Mel'čuk and Polguère (2009) postulates three types of syntagmatic dependency relations, namely semantic dependency, syntactic dependency and morphological dependency. The only feature that

every dependency-based model shares is that there is no abstract intermediate node, such as *VP* (Verbal Phrase), to constitute the final phrase, unlike all Chomskyan-based grammars. In this sense, the adpositional paradigm is similar to dependency grammars. However, in all these models, prepositions are treated as modifiers of nouns, i.e., nouns are governors, prepositions are dependents, exactly as if prepositions were adjectives—in our perspective, a fatal error. Therefore, we consider dependency in a different way inside constructive adpositional grammars. It is worth noticing that Tesnière (1959) never considered his grammar based on the notion of dependency, preferring rather the more general term *connexion* ('connection', in French). In fact, he always called his own model 'structural syntax'.

[12] In the formal model depicted in Appendix B, we will reserve the term 'valency' for the verbal constructions; instead we will use the more general concept of grammar product to represent the structure of actants.

[13] One could indeed compare the verb to a kind of **crossed atom**, which can attract a number more or less high of actants, in proportion to the number more or less high of hooks needed to maintain the respective dependencies (our translation).

[14] Please note that in cognitive linguistics the term 'valency' is used in a different way—see the glossary in Taylor (2002) for that.

[15] Please note that we are not speaking about 'intransitive' or 'transitive' verbs, but instead of 'monovalent' or 'bivalent' verbs. Unfortunately, the traditional account of valency did not learn the main lesson from linguistic typology, i.e., that the traditional dichotomy 'subject/object' is too poor to give an account of nominative-accusative *vs.* ergative-absolutive verb classes (Dixon, 1994).

[16] The wording 'passively' is tentative: a more precise account of this difference will be clear throughout the next two chapters.

[17] The construction Y *sold* is a derived one from X *sell* Y *to* Z *for* W where all actants save the SOLD OBJECTIVE (Y) are hidden.

[18] Please note that our treatment of the example is different in many aspects. Fillmore (2003b) collects his classic papers on semantic roles, written between 1969 and 1978. In the sequel, we will refer to these papers in this collection for the reader's sake.

[19] The reader should notice that examples (2-3hj) can be well-formed phrases, but the point is that they carry a different meaning than (2-3g), describing a distinct scene.

[20] For example, Croft (2009) shows how the EAT frame differs in English and in German, where in the latter there is a distinction between A-EAT for animals (*fressen*) and H-EAT for humans (*essen*) which in English is defined otherwise. Semantic frames and the analysis of constructions can be fruitfully put into relation; in this book, the focus remains the constructions while the semantic frames are put into the background.

[21] We will come back on this crucial point in the next chapters.

[22] In the formal model, transformations are responsible for explaining how complex constructions take place, e.g., the active-to-passive transformations. In fact, constructions of transformed expressions are still generated by some appropriate constructions, which are synthetically described as the image of a transformation. This is possible since constructions are indexed families of arrows and transformations are functors, so they also map arrows to arrows, eventually leading to map constructions into constructions.

[23] For an analysis of the construction of English dative alternation, see at least Goldberg (1995). Dative alternation in Dutch is explained, in comparison and contrast to English, in Chapter 3 of van der Beek (2005), while Abreu Gomes (2003) explores Brazilian Portuguese, with a reference to Tsunga for comparison. We do not pursue a comprehensive treatment of the phenomenon of dative alternation—that would occupy at least a full chapter dedicated to it—but rather how transformation works in real-case constructions.

[24] In fact, pronouns, being only placeholders for nouns, are usually not prominent, as the speaker's non-locutionary assumption is that the listener knows who or what the pronoun is referring to.

[25] Technically, the adtree on the right of Figure 2.20 is a conjugate construction of the one on the left, where the second actant is exchanged with the first one. Formally, the two trees are equivalent, except for the different content of the second actant, but the representation provides an account for emphasis and collocation on the syntagmatic axis, see Appendix B for the details.

[26] Notice how a conjugate construction has been used, exchanging the second and the third actants, to represent the different collocations of morphemes on the syntagmatic axis. This holds also in Example (2-7).

Chapter 3

[1] We limit ourselves, unless otherwise stated, to morphological analysis in synchrony. For example, in Italian the word *giornata* (day-having) is synchronically made by two morphemes, i.e., *giornat-* (day-having) and *-a* (singular and feminine), even if diachronically *giornat-* could be further analysed as *giorn-* (day) and *-at-* (past participle). This kind of discourses, really interesting for the analysis of a single natural language, is not central for our purposes. A more precise and formal treatment of morphemes and their components will be explained in the next chapters.

[2] See for instance Bozsahin (2002) for a computational treatment of grammar starting from morphemes instead of words.

[3] We will use the term 'class' in a naive sense, i.e., equivalent to a labelled set (also 'set' in a naive sense), while the term 'category' will be mostly used in a mathematical, strong sense. Sometimes the term 'category' will be used linguistically, i.e., equivalent to

a definite, predetermined class. The context will clarify which sense is intended; if not, the reader will be helped with a proper specification in order to disambiguate.

[4] Moreover, there is an additional problem in using traditional grammatical categories like 'noun', 'verb', 'adjective', 'adverb', 'passive voice', i.e., these categories are usually considered 'fixed into the morpheme', because they derived from classical tradition, i.e., from ancient Greek and Latin grammarians. As we will see in the sequel, we consider grammar categories as *functions* of grammar characters instantiated into morphemes.

[5] As far as we know, besides Gobbo (2009), only Miner (2008) uses Whorf (1945).

[6] The term 'stative' is derived from 'stativation' as used by Whorf (1945), while 'verbant' is a term introduced in the sense we use here by Hagège (1996).

[7] We retain the linguistic use of the term as the default meaning, while we will use the term *adjunction* in the technical mathematical sense. The context will help the reader to avoid ambiguity; if not, we will explicit its use.

[8] In the majority of constructions belonging to Standard Average European languages character raising is the norm, while languages belonging to other typological families show a considerable amount of character imposition.

[9] We still do not have examples for some of them, e.g., generic relation circumstantial-verbant. Nonetheless, the formal model shows that they are possible, at least in theory. The convention for the names is simple: the first term is the dependent, the second one the governor.

[10] Of course, marked nominative is not unknown; nonetheless, it is not the default case in typological terms (Dixon, 1994, 63–67).

[11] Blake (1990) speaks of 'detransitivity' about the valency reduction. We prefer not to use the term 'transitivity' as it covers too many cases of information prominence, preferring to use the expressions 'valency decrease' or 'valency increase'.

[12] The transformation of the grammar character I > A > Y in Figure 3.12 refers to the internal structure of the word 'sudato'.

[13] Of course, holophrases (12 months), which have only one element, are really pre-morphosyntactic, as explained by Tomasello (2003).

[14] English examples are reported in Tomasello, while the Italian ones are transcriptions of Leonardo's speech (19 months), Federico Gobbo's son.

[15] In constructive linguistics, the meaning of the parts is the domain of semantics, while the meaning-in-use, i.e., the significance of the whole, is the domain of pragmatics. We will come again on this crucial distinction.

[16] Unfortunately, traditional grammarians of Latin introduced a very confusing terminology, namely 'subjective genitive' (the father being the subject of love, i.e., with dependency) and 'objective genitive' (the father being the object of love, i.e., with government). Our proposal, being abstract and of general validity, has the greater advantage not to be bound to a particular grammar tradition. Our guideline is simple: we retain the established terminology if and only if it is not ambiguous in constructive linguistic terms; otherwise, we introduce novel, non-ambiguous terms.

[17] We use the word 'class' in a relaxed sense, meaning a named set, while a *linguistic* category is a *defined* class, such as 'English personal pronouns'. Of course, a linguistic category has sense only within a particular language. For instance, if we speak about personal pronouns within the Standard Average European Sprachbund the object we are referring to is always a class, not a linguistic category.

[18] The word 'quantifier' is used in the linguistic sense, non the logical one, unless otherwise specified.

[19] There is often a misunderstanding about determiners: this class of words have great consequences at a pragmatic level (participant's beliefs being a considerable part of pragmatic information), nonetheless they are often analysed at a semantic level. We consider that the analyses of determiners as semantic words follow a wrong perspective.

[20] In the grammar tradition of German, there is even a mnemonic to memorise the main circumstances: the word *tekamolo* means *Temporal* (temporal), *Kausal* (causal), *Modal* (modal) and *Lokativ* (locative).

[21] From a diachronically point of view, the French stative *paysage* is a grammaticalization of *pays* and *-age*. In this book, we won't delve in this kind of grammaticalization phenomena, which are left as a further work.

[22] Although the formal model allows for a more general notion of transformation, which is useful and largely employed in the text, the synthesis of complex constructions is the principal motivation behind the introduction of transformations.

Chapter 4

[1] On the other hand, sometimes very distant typological languages can share very similar transformation patterns. The relation between typological proximity in every pair of languages and their respective transformation patterns for the same phenomenon is beyond the scope of the present book.

[2] It is impossible to give an exhaustive coverage of morphology, even within a single language: what we aim to do in this book, is to explain how to represent within the adpositional paradigm some relevant and well known morphologic phenomena, in particular grammar character change obtained by derivation—for example, how the English

adjective *well* is transformed into the abstract noun *wellness*. This level of analysis will be explained in Chapter 5.

[3] We mean final governors and dependents, i.e., the ones not further expandable. In practice, leaves *without* the triangles ($\triangle$). For these reasons, it seems feasible to use two different symbols to sign empty adpositions (ϵ) and void governors and dependents ($\square$).

[4] This is not a real transformation, but it helps the reader to understand every compact notation we use throughout the book, therefore we decided to explain this point here.

[5] It is quite obvious that both governor and dependent cannot be void at the same time: if so, the adposition is not an adposition, but only a leaf and therefore it should be considered as a governor.

[6] From a formal point of view, information prominence is only an adtree decoration—see Appendix B for details.

[7] We adopt the law of parsimony (Occam's razor) in this respect: make explicit only the sememes you really need for your purposes, in this case the explanation of redundancy.

[8] In the formal model, sememes are just attributes of morphemes, i.e., of the objects of the morphemic category. Although it is convenient for the exposition, sememes do not form a hierarchy, so speaking of subordinate sememes is somewhat improper. In fact, they are flattened in the formal representation.

[9] As this volume deals with the foundations of the constructive adpositional paradigm, we will not deal in detail with a typological survey, which is of undoubtful interest, but out of our purposes here.

[10] We take as the reference grammar of English Alexander (1988), with some further examples useful for our purposes.

[11] For this reason, constructions such as *will be* or *shall be* should be listed as constructions of *will* and *shall* respectively. The imperative constructions in (4-3cd) are active forms of *to be*.

[12] Pragmatically, in (4-3a) the main point is that Liza is quiet, not that is a girl. If we omit the adjective, obtaining *Liza is a girl*, the immediate reaction is: so what? There is something lacking, exactly the information whose vehicle is the adjective.

[13] We have already seen an example of this fact in (1-3)—see again Figure 1.2. In fact, in *È la rossa di Maranello* the stative is the *second* actant, which is prominent, not the first one.

[14] With 'complex constructions' we mean informally constructions with more than one verbant, i.e., having only one governor with the grammar character I, in contrast with 'simple constructions', where only a verbant is found.

[15] If we have a trivalent verb, like in *Paul is giving lessons to Mary*, the adtree having *give* as the governor would be lengthened appropriately for the third actant Z_{mean}. This adtree is left to the reader.

[16] This example is, probably, the most complex in the whole book: this happens because it interleaves constructions at the morphological level with constructions on the linguistic level, mixing them together to convey the intended meaning. In fact, *going to* is usually considered a completely grammaticalized construction called the 'present progressive tense'; our analysis shows that it is not special, but rather a complex instance of the infinitive, the gerund and the *esse* constructions.

[17] This construction is also found in Italian as well: *Paolo è preparato* corresponds to (4-4e) both syntactically and semantically. We guess that exactly the same construction can be found in a consistent number of languages belonging to the Standard Average European Sprachbund.

[18] These examples by Horn (1980) are quoted by Blake in his volume, which summarises the main results of the research made within the Relational Grammar paradigm (Blake, 1990, section 2.4:31). Here, they have been adapted to the adpositional paradigm. We have chosen not to give precise names to these classes of bivalent English verbs, because every possible name we found clashed with some tradition in English linguistics, and this book is not focused on the English adpositional grammar.

[19] The idiomatic expression *there once was a time...*, although it seems a *there*-construction, is really governed by *once*: in fact, *once was a time...* is a synonym, while *there was a time...* is not.

[20] It is possible that *esse*-constructions can be reconducted to this class indeed: compare (4-34ab) to *he became (someone) made* and *he is (someone) mad*. However, this line of reasoning is left as a further work, as it is out of the foundational scope of this volume.

[21] There is a last argumentation question posed by Cicero, called *facultas*, which answers the question "which instruments are you using in your argumentation?". This was omitted for brevity, but this omission does not change our line of reasoning.

[22] Quadrivalent verbs are rare: in fact, before the concept of money, languages did express commercial transaction without PRICE, using verbs like *donate*

[23] Actually, the same phrase may be represented by a different adtree that uses the infinitive construction to build the in-valent actants. This representation is correct, but it disregards the causative sememe which is present in the verbant governing the phrase. Our point here is that there is a proper construction that takes place because of the presence of the causative sememe. The alternative construction has a slightly different meaning, in that it does not convey the 'forcing' of the dependent actions by the governor.

Chapter 5

[1] Phonology is the most language-bound level of analysis, and therefore it is not central for the purposes of this book, while pragmatics will be addressed in the next chapter.

[2] Within the cognitive grammar approach, only Wierzbicka (1988) does not follow the pan-encyclopædic approach.

[3] The choice of the phrasal construction has semantic as well as pragmatic consequences. The part concerning pragmatics will be explained in the next chapter.

[4] In philosophical terms, we endorse the conceptualist approach. Unlike Dewey, Wittgenstein and Quine, there *is* more to the meaning of an expression than the overt use that we make of the expression, paraphrasing Quine (1987, 130) in Taylor (2002, 63). In other words, language cannot be reduced to a mere observation of behaviours: human beings are more intelligent than ELIZA (Weizenbaum, 1966).

[5] We do not deny the relevance of encyclopædic knowledge: simply, we are deeply convinced that language should be described primarily within its coherence as a system, instead of giving account of the symbol grounding and a particular world model, which is always questionable. In respect to world models, we are agnostic.

[6] From the point of view of a lexical semanticist, we follow the tradition of monosemic analysis, begun at least from De Saussure (1970). Adherents of the polysemic analysis, like Katz and Fodor (1963), prefer to treat cases like 'bachelor' as a continuum, where there is a prototypical abstract meaning like 'unfulfilled in typical male role'—see Kornai (2008, 488-489) for details. In our perspective, when performing human or machine translation, such a perspective would further increase the already complex problem of transferring semantics from one language to another. On the contrary, from a monosemic perspective, the fact that these morphemes are disjoint greatly simplifies the task. For example, in Italian, $bachelor_1$ is rendered with *celibe*, while $bachelor_3$ with *diplomato*.

[7] The problem of the correspondence of sememes and their configuration within morphemes and constructions belonging to different languages, even if an important and challenging problem, is beyond the scope of the present book.

[8] As the reader, after the previous chapters, may guess, adpositions are the fundamental constituent of the functional words. But, although functional words are always related to adpositions, there are elements of this class not belonging to the set of adpositions, e.g., pronouns or correlatives.

[9] Here and in the following, we use the word 'constructive' in a general sense, not in the strictly mathematical one. Also, the notion of 'open set' is used in a linguistic context and should not be confused with the technical meaning of mathematical open set which refers to topology. Here, open set means a set that is not fixed and that can vary in time.

[10] In developing his 'structural syntax'—unfortunately still not available in English Tesnière (1959) devotes Part 3 (approximately 400 pages) entirely to transference. A partial translation of Tesnière (1959) is available in Italian by Proverbio and Trocini Cerrina (2001), but not the one devoted to transference. No other translations are available in the literature, to our knowledge. We would appreciate any different information by readers in such respect.

[11] Tesnière did distinguish: the *transférende* ('transfer-worth'), i.e., where transfer starts; the *transféré* ('transferred'), the result of the transfer act; the *translatif* ('transferrer'), the "morphological marker for transference" (Tesnière, 1959, 367, our translation). Unfortunately, he did not generalise accordingly this triple, letting transferrers (i.e., our adpositions) exist only if morphologically marked. In our view, this is one of the main limit of its model.

[12] It is worth remarking that the constructions devoted to build 'words' are distinct from constructions devoted to the inter- on intra-phrasal level. Although they share the same structure and the same principles, and, to some extent, they operate by interleaving—see, e.g., the correlatives—morphological constructions are devoted to construct the 'atoms' for the linguistic level.

[13] Words denoting number are puzzling: in fact, as *two* is a lexeme, *two-hundred* is the result of a special construction, although it operates as a lexeme. In technical contexts, it is even accepted the *3-fold* form, e.g., topology.

[14] The case of language policy and planning of planned languages, such as Esperanto, is different. The interested reader can see at least Gobbo (2008).

[15] But it is easy to design an artificial language which is completely suppletive. In this respect, artificial languages are of interest, since they provide pathological contrastive examples. Similarly, it is easy to design artificial languages where suppletion is absent, e.g., most computer programming languages are of this sort. Please note that planned languages like Esperanto are Quasi-Natural Languages (QNL), not artificial (Lyons, 2006, 69–70).

[16] Strictly speaking, a transfer is a transformation with a rigid pattern. But, for the moment being, we will ignore this detail, as transfers can be represented as transformations, as explained in the following.

[17] Notice how synonyms can be retrieved by querying for a set of sememes. So our dictionary is also a thesaurus thanks to the information content of sememes.

[18] Here and in the following, representation means the way a piece of data appears in the database as seen by the user, and not its internal representation which relies on the technicalities of database technology.

[19] Representing an abstract construction as a query is equivalent to say that an abstract construction denotes the set of all its possible instances, which, in turn, are the instances of the general constructions composing it.

[20] There are exceptions, of course, due to the fact that grammaticalization in not uniform. For example, the past tense of 'go' is 'went', see Section 5.2 under suppletion, and so a transformation has to plan for a change of morphemes where the new morph is not simply derived from the old one, e.g., appending a '-ed'. But this effects can be obtained via composition of simpler transformations where only the last step supports suppletion. But the computational analysis of the adpositional paradigm is far beyond the scope of the present text.

[21] Of course, as usual when inferring general facts from empirical analyses, in the lack of a stronger argument, we warn the reader that our conclusions, although reasonable, have not been extensively tested on large corpora of examples from languages of all over the world.

Chapter 6

[1] Grammaticalization is not free from the community of the language speakers: in other words, they can influence the language itself, either consciously by language policy and planning, or not, by *lassez faire*.

[2] It may be that a Universal Grammar—as put by Chomsky and generativists in general—evolved during phylogeny of the human species; unfortunately, we do not know how language originated in human evolution (Tomasello, 2003). That is why we compare grammaticalization within different languages, in order to extract common, general patterns of use, i.e., constructions.

[3] Only human beings are potentially sign producers and receivers, as far as it known by now, and this switching capability lead to the property called 'bidirectionality of the sign' as stated by Saussure (Tomasello, 2003).

[4] In particular, a seminar was organised in order to explore the possibilities of further research in the area of constructive conversation analysis, taking transcripts of conversations between therapists and clients as the raw language data. The result of this collaboration was published in a book in Italian (Gobbo, 2011).

[5] The Senior Researcher and Director of Training of the European Center for Therapeutic Assessment is Stephen E. Finn, already founder of the Center for Therapeutic Assessment at Austin, Texas, where he is a licensed clinical psychologist, and Clinical Assistant Professor of Psychology at the University of Texas.

[6] As a first step, an exploratory Master's thesis is currently devoted to this research program, written in Italian by Riccardo Preziosi.

[7] More precisely, (6-1a) is extracted from Darvill-Evans' imaginative book *Deceit* (1993), while (6-1b) is part of two conversations recorded by 'Dorothy' (PS087) between 13 and 20 March 1992 with twentyfive interlocutors. Finally, (6-1c) is part of 28 conversations recorded by 'Arthur2' (PS50T) between 27 February and 2 March 1992 with seven interlocutors.

[8] We use the expression 'mental state' naively here, as a discourse on the ontological status of the mind and its states is out of the scope of this book. However, as it will be clear, the philosophical perspective of realism (Searle's mind-in-brain) perfectly fits our line of reasoning (Searle, 2004, for instance).

[9] In the literature, there are two approaches in order to formalise pragmatic information embedded in the locution, which can be regarded as formalisations of pragmatic vagueness. The first approach start with Grice (1975), who advocates a theory of scalar implicatures ruled by a set of conversational principles generally valid. Unfortunately, there is no agreement among scholars about the content, number and validity of the conversational principles themselves. The second approach puts the problem in the real of grammar: there is an algorithm that locally computes the implicatures that can be inferred from the locution (Chierchia, 2004, for instance). The disputable point here is in the locality of the implications. However, not only judgement is divided but even experimental evidence is lacking. For a recent analysis on this topic, see Ippolito (2011).

[10] The lambda symbol comes from λ-calculus since its characterising operation is abstraction. The idea is to abstract the illocutionary information from the linguistic, locutionary part, which is the expressed one.

[11] Broadly, we have a turn change when the roles of speaker and listener are switched among the actors of the conversation. For our purposes, we do not need any further refinement of the concept of turn and turn change.

[12] Searle's work can be traced at least from Searle (1969), when he proposed an alternative taxonomy to Austin's, until Searle (2010), more focused on the social consequences of his philosophy of language.

[13] This depict is quite natural not only in Western philosophy, but also in other cultures of the world, such as Chinese, where the human being is considered the axis that glues together the sky (i.e., the mind) and the earth (i.e., the world).

[14] Besides what already quoted from Searle (2010) and Searle (2002), other relevant sources are reflections on the intentionality of seeing, believing, desiring and remembering in Searle (1983).

[15] This is one of Searle's example (Searle, 1983, 33)

[16] It is possible that before agriculture human beings share commonly this role; anyhow, it is indubitable that the expression of feelings by linguistic means is a relevant part of the human species specificity.

[17] In order to mark the special role of Declarations, they are indicated with a special capital letter, in symbols and in text.

[18] There was a second example taken from the description made by Harry, another client, during a Rorschach test (Finn, 2007, 77). Unfortunately, without conversational turns, with a clear task of picture description, like a Rorschach, most turns are simply assertives, with some powerful expressive, like *I sense a hesitance in me to accept this idea that this bat might be in trouble.* However, from a constructional point of view, such sample is far less informative.

Chapter 7

[1] Another perspective could be for instance the analysis of politeness in constructive terms (Brown and Levinson, 1987) or dialogue in terms of the Mixed Game Model (Weigand, 2011).

Appendix B

[1] Notice that we defined conjugates only for construction from a grammar character. Here, we suppose η to be a component of such a construction θ, and the conjugate η^* is the obviously chosen component of some conjugate of θ.

[2] Although it is possible and not essentially more difficult to consider general grammar categories, understanding the meaning of definitions becomes more complex as adtrees are not unique modulo conjugates.

[3] Actually, the formal reader would notice that the statement is not perfectly correct: we don't meet the condition saying that every construction can be described as the composition of constructions of degree 1. If we define *pseudo-grammar categories* as grammar categories except for the preceding condition, which gets relaxed into requiring the degree to be at most 1, then $\overline{\mathbb{G}}$ becomes a pseudo-grammar category. Essentially, the only point we loose here is the binary adtree representation of objects; instead, we get that objects are represented by adtrees having one or two sons in every internal node.

[4] We avoided a technical statement because some care should be taken when dealing with empty strings and symbol. The statement holds when the empty symbol Λ is explicit in the strings.

[5] Again, we are assuming explicit Λ symbols. With implicit empty symbols, a slightly more complex definition must be given.

Bibliography

Christina Abreu Gomes, 2003. "Dative alternation in Brazilian Portuguese: typology and constraints." *Language Design* 5: 67–78

Kazimierz Ajdukiewicz, 1935. "Die syntaktische Konnexität." *Studia Philosophica* 1: 1–27

Louis G. Alexander, 1988. *Longman English Grammar*. Harlow: Longman

Randy Allen Harris, 2010. "Chomsky's other Revolution." In Kibbee (2010), pp. 237–264

John Langshaw Austin, 1962. *How to Do Things With Words*. Cambridge, MA. and London: Harvard University Press

Franz Baader and Jörg H. Siekmann, 1994. "Unification theory." In Dov M. Gabbay, C. J. Hogger, and J. A. Robinson, editors, "Handbook of Logic in Artificial Intelligence and Logic Programming," pp. 41–125. Oxford: Oxford University Press

Mark Baker, 1997. "Thematic Roles and Syntactic Structures." In Haegeman (1997), chapter 2, pp. 73–137

Yehoshua Bar-Hillel, 1953. "A quasi-arithmetical notation for syntactic description." *Language* 29: 47–58

Leonoor Johanneke van der Beek, 2005. *Topics in Corpus-Based Dutch Syntax*. Ph.D. thesis, Doctoraat in de Letteren aan de Rijksuniversiteit Groningen

Johan F. A. K. van Benthem, 1995. *Language in Action: Categories, Lambdas, and Dynamic Logic*. Cambridge, MA.: MIT Press

Barry J. Blake, 1990. *Relational grammar*. London: Routledge

Thomas P. Bonfiglio, 2010. *Mother tongues and nations: the invention of the native speaker*. Berlin & New York: Mouton de Gruyter

Pier Giorgio Borbone, Alessandro Mengozzi, and Mauro Tosco, editors, 2006. *Loquentes linguis. Studi linguistici e orientali in onore di/ Linguistic and Oriental Studies in Honour of/ Lingvistikaj kaj orientaj studoj honore al Fabrizio A. Pennacchietti*. Wiesbaden: Harassowitz Verlag

Francis Borceux, 1994. *Handbook of Categorical Algebra I*, volume 50 of *Encyclopedia of Mathematics and its Applications*. Cambridge: Cambridge University Press

Cem Bozsahin, 2002. "The Combinatory Morphemic Lexicon." *Computational Linguistics* 28 (2): 145–186

Douglas Bridges and Fred Richman, 1987. *Varieties of Constructive Mathematics*, volume 97 of *London Mathematical Society, Lecture Notes Series*. Cambridge: Cambridge University Press

Viggo Brøndal, 1940. *Præpositionernes Theori*. Copenhagen: Ejnar Munksgaard

Penelope Brown and Stephen C. Levinson, 1987. *Politeness: Some universals in language usage*. Cambridge: Cambridge University Press

Silvio Ceccato, editor, 1961. *Linguistic Analysis and Programming for Mechanical Translation*. Milano / New York: Feltrinelli editore / Gordon & Breach Publisher

Isabella Chiari, 2002. *Ridondanza e linguaggio: un principio costitutivo delle lingue*. Roma: Carocci

Gennaro Chierchia, 2004. "Scalar implicatures, polarity phenomena, and the syntax/pragmatics interface." In Adriana Belletti, editor, "Structures and Beyond," pp. 39–103. Oxford: Oxford University Press

Noam Chomsky, 1957. *Syntactic Structures*. The Hague: Mouton

Noam Chomsky, 1965. *Aspects of the Theory of Syntax*. Cambridge, MA.: MIT Press

Noam Chomsky, 1981. *Lectures on Government and Binding*. Dortrecht: Foris

Noam Chomsky, 1992. *A Minimalist Program for Linguistic Theory*. Cambridge, MA.: MIT Press

Alonzo Church, 1940. "A formulation of the simple theory of types." *Journal of Symbolic Logic* 5: 56–68

Bertrand Comrie, 1989. *Language Universals and Linguistic Typology*. Chicago: Chicago University Press

Thierry Coquand, 1996. "An Algorithm for Type-checking Dependent Types." *Science of Computer Programming* 26 (1–3): 167–177

Greville G. Corbett, 2000. *Number*. Cambridge textbooks in linguistics. Cambridge: Cambridge University Press

William Croft, 2001. *Radical Constructions Grammar: Syntactic Theory in Typological Perspective*. Oxford: Oxford University Press

William Croft, 2009. "Connecting frames and constructions." *Constructions and Frames* 1 (1): 7–28

Ferdinand De Saussure, 1970. *Introduzione al secondo corso di linguistica generale (1908 – 1909)*. Roma: Astrolabio. Editor: Robert Godel. Italian edition: Raffaele Simone

Ralph Debusmann, 2001. *A declarative grammar formalism for dependency grammar*. Ph.D. thesis, Universität des Saarlandes

Sharon J. Derry, Morton Ann Gernsbacher, and Christian D. Schunn, editors, 2005. *Interdisciplinary collaboration: an emerging cognitive science*. Lawrence Erlbaum Associates: Mahwah, N.J.

Robert M. W. Dixon, 1994. *Ergativity*. Cambridge: Cambridge University Press

Germain Dondelinger, Francoise Madray-Lesigne, and Jeannine Richard-Zappella, editors, 1995. *Lucien Tesnière aujourd'hui*. Peeters Publishers

Sabine Fiedler, 1999. *Plansprache und Phraseologie. Empirische Untersuchungen zu reproduziertem Sprachmaterial im Esperanto*. Frankfurt am Main: Peter Lang

Sabine Fiedler, 2002. *Esperanta frazeologio*. Rotterdam: Universala Esperanto-Asocio

Charles J. Fillmore, 1982. "Frame Semantics." In Geeraerts et al. (2006), chapter 10, pp. 373–400. Originally published in 1982

Charles J. Fillmore, 2003a. *The Case for Case*, chapter 2, pp. 23–122. Volume 1 of Fillmore (2003b)

Charles J. Fillmore, editor, 2003b. *Form and Meaning in Language: Papers on Semantic Roles*, volume 1. Stanford, CA.: CSLI Publications

Charles J. Fillmore, 2003c. *Toward a Modern Theory of Case*, chapter 1, pp. 1–21. Volume 1 of Fillmore (2003b)

Stephen E. Finn, 2007. *In Our Clients' Shoes: Theory and Techniques of Therapeutic Assessment*. Lawrence Erlbaum Associates: Mahwah, N.J.

Dov M. Gabbay, 1996. *Labelled Deductive Systems, volume 1*, volume 35 of *Oxford Logic Guides*. Oxford: Oxford University Press

Michael R. Garey and Davis S. Johnson, 1979. *Computer and Intractability: A Guide to the Theory of NP-Completeness*. New York: W. H. Freeman and Company

Dirk Geeraerts, René Driven, and John R. Taylor, editors, 2006. *Cognitive linguistics: basic readings*. Berlin & New York: Mouton de Gruyter

Federico Gobbo, 2008. "Planning languages and language planning: The contribution of Interlinguistics to cross-cultural communication." In Franks Boers, Jeroen Darquennes, and Rita Temmerman, editors, "Multilingualism & Applied Comparative Linguistics," volume 2, pp. 24–60. Newcastle upon Tyne: Cambridge Scholars Publishing

Federico Gobbo, 2009. *Adpositional Grammars: A multilingual grammar formalism for NLP*. Ph.D. thesis, University of Insubria, Varese, Italy

Federico Gobbo, 2010. "Fondamenti delle grammatiche adposizionali / Fundamentoj de la adpoziciaj gramatikoj." *InKoj. Interlingvistikaj Kajeroj* 1 (1): 35–63

Federico Gobbo, 2011. *Nei panni delle nostre conversazioni terapeutiche: analisi conversazionale costruttiva.* Milano: Raffaello Cortina

Adele E. Goldberg, 1995. *Constructions: A Construction Grammar Approach to Argument Structure.* Chicago: Chicago University Press

Adele E. Goldberg, 2006. *Constructions at work.* Oxford: Oxford University Press

Algirdas J. Greimas, 1966. *Structural Semantics: An Attempt at a Method.* Nebraska: University of Nebraska Press

Paul Grice, 1975. "Logic and conversation." In Peter Cole and Jerry L. Morgan, editors, "Syntax and Semantics 3: Speech Acts," pp. 39–103. New York: Academic Press

Liliane M. V. Haegeman, editor, 1997. *Elements of grammar: handbook in generative syntax.* Dordrecht: Kluwer Academic Publishers

Claude Hagège, 1996. *L'homme de paroles: contribution linguistique aux sciences humaines.* Paris: Fayard

Randy Allen Harris, 1995. *The Linguistic Wars.* Oxford: Oxford University Press

Martin Haspelmath, 2001. "The European linguistic area: Standard Average European." In Martin Haspelmath, Wulf Oesterreicher, and Wolfgang Raible, editors, "Handbuch der Sprach- und Kommunikationswissenschaft," volume 20.2, pp. 1492–1510. Berlin & New York: Mouton de Gruyter

Jaakko Hintikka, Julius Moravcsik, and Patrick Suppes, editors, 1974. *Approaches to Natural Language: proceedings of the 1970 Stanford workshop on Grammar and Semantics.* Reidel

Kouichi Hirata, Keizo Yamada, and Masateru Harao, 1999. "Tractable and Intractable Second-Order Matching Problems." In Takano Asano, Hideki Imai, D. Lee, Shinichi Nakano, and Takeshi Tokuyama, editors, "Computing and Combinatorics," volume 1627 of *Lecture Notes in Computer Science*, pp. 432–441. Frankfurt Am Main: Springer

Laurence Horn, 1980. "Affixation and the unaccusative Hypothesis." *CLS* 16: 134–146

Richard Hudson, 2007. *Language Networks: The New Word Grammar.* Oxford: Oxford University Press

Michela Ippolito, 2011. "A Note on Embedded Implicatures and Counterfactual Presuppositions." *Journal of Semantics* 28: 267–278

Peter Johnstone, 2002. *Sketches of an Elephant: A Topos Theory Compendium, Volume 2*, volume 44 of *Oxford Logic Guides*. Oxford: Oxford University Press

Jerome Kagan, 2009. *The Three Cultures: Natural Sciences, Social Sciences, and the Humanities in the 21st Century*. Cambridge: Cambridge University Press

Jerrold J. Katz and Jerry A. Fodor, 1963. "The structure of a semantic theory." *Language* 39: 170–210

Ruth Kempson, Wilfried Meyer-Viol, and Dov Gabbay, 2001. *Dynamic Syntax: The Flow of Language Understanding*. London: Blackwell

Douglas A. Kibbee, editor, 2010. *Chomskyan (R)evolutions*. Amsterdam: John Benjamins

Pekka Kilpeläinen, 1992. *Tree Matching Problems with Applications to Structured Text Databases*. Ph.D. thesis, Department of Computer Science, University of Helsinki. Available as Report A-1992-6, Helsinki, Finland, November 1992

András Kornai, 2008. "On the proper definition of information." In Terrell Ward Bynum, Mariacarla Calzarossa, Ivo de Lotto, and Simon Rogerson, editors, "Living, Working and Learning Beyond Technology," pp. 488–495. Mantua: University of Pavia

George Lakoff, 1997. *Women, Fire and Dangerous Things: What Categories Reveal about the Mind*. Chicago: Chicago University Press. Original edition 1990

Joachim Lambek, 1958. "The mathematics of sentence structure." *American Mathematical Monthly* 65: 154–170

Joachim Lambek and P.J. Scott, 1986. *Introduction to Higher Order Categorical Logic*, volume 7 of *Cambridge Studies in Advanced Mathematics*. Cambridge: Cambridge University Press

Ronald W. Langacker, 1987. *Foundations of Cognitive Grammar*, volume 1. Stanford University Press. Theoretical Prerequisites

Ronald W. Langacker, 1990. *Concept, Image and Symbol: The Cognitive Basis of Grammar*. Berlin & New York: Mouton de Gruyter

Ronald W. Langacker, 1991. *Foundations of Cognitive Grammar*, volume 2. Stanford University Press. Theoretical Prerequisites

Ronald W. Langacker, 1995. "Structural syntax: the view from cognitive grammar." In Dondelinger et al. (1995), pp. 13–37

John Lyons, 2006. *Natural language and universal grammar. Essays in linguistic theory*, volume 1. Cambridge: Cambridge University Press

Saunders Mac Lane, 1998. *Categories for the Working Mathematician*, volume 5 of *Graduate Texts in Mathematics*. Frankfurt Am Main: Springer, 2nd edition

Saunders Mac Lane and Ieke Moerdijk, 1992. *Sheaves in Geometry and Logic*. Universitext. Frankfurt Am Main: Springer

Ana Maria Madeira, 1994. "On the Portuguese inflected infinitive." *UCL Working Papers in Linguistics* 6

Christopher D. Manning and Hinrich Schütze, 1999. *Foundations of Statistical Natural Language Processing*. Cambridge, MA.: MIT Press

John C. Martin, 1997. *Introduction to Languages and the Theory of Computation*. Programming Languages Series. McGraw Hill, second edition

Per Martin Löf, 1984. *Intuitionistic Type Theory*. Napoli: Bibliopolis

Peter Hugoe Matthews, 1993. *Grammatical theory in the United States from Bloomfield to Chomsky*. Cambridge: Cambridge University Press

Igor Mel'čuk and Alain Polguère, 2009. *Dependency in Linguistic Description*. Amsterdam: John Benjamins

Ken Miner, 2008. "Noto pri esperantaj radikoj." *Lingva Kritiko*

Richard Montague, 1973. "The Proper Treatment of Quantification in Ordinary English." In Hintikka et al. (1974), pp. 221–242

Andrea Moro, 2010. *Breve storia del verbo essere*. Milano: Adelphi

Glyn Morrill, 2007. "A Chronicle of Type Logical Grammar: 1935–1994." *Research on Language & Computation* 5: 359–386. ISSN 1570-7075. URL `http://dx.doi.org/10.1007/s11168-007-9034-2`

Joakim Nivre, 2005. "Dependency grammar and dependency parsing." Technical report, Växjö University

Fabrizio A. Pennacchietti, 1976. "La prepozicia sistemo de Esperanto." *Esperantologiaj Kajeroj I* pp. 137–153

Fabrizio A. Pennacchietti, 2006. "Come classificare le preposizioni? Una nuova proposta." *Quaderni del Laboratorio di Linguistica* 6: 1–18

Fabrizio A. Pennacchietti, 2009. "The Prepositional System of Classic Syriac and that of Sureth." In Shamuel (2009), pp. 35–49

Mati Pentus, 1993. "Lambek Grammars are Context Free." In "Proceedings of Eighth Annual IEEE Symposium on Logic in Computer Science," pp. 429–433. IEEE

Chaïm Perelman and Lucie Olbrechts-Tyteca, 1958. *Traité de l'argumentation. La nouvelle rhétorique*. Bruxelles: Editions de l'Université de Bruxelles

David M. Perlmutter, 1978. "Impersonal passives and the unaccusative hypothesis." *BLS* 4: 157–189

David M. Perlmutter and Paul M. Postal, 1984. "The 1-Advancement Exclusiveness Law." In Perlmutter and Rosen (1984), pp. 81–125

David M. Perlmutter and Carol G. Rosen, editors, 1984. *Studies in Relational Grammar II*. Chicago: Chicago University Press

Germano Proverbio and Anna Trocini Cerrina, 2001. *Elementi di sintassi strutturale*. Torino: Rosenberg & Sellier

Willard van Orman Quine, 1987. *Quiddities: An Intermittently Philosophical Dictionary*. Cambridge, MA: Harvard University Press

Aarne Ranta, 2004. "Grammatical Framework: A Type-theoretical Grammar Formalism." *The Journal of Functional Programming* 14 (2): 145–189

Aarne Ranta, 2009. "GF: A Multilingual Grammar Formalism." *Language and Linguistics Compass* 3 (5): 1242–1265

Giovanni Sambin and Silvio Valentini, 1998. "Building a Toolbox for Martin-Löf's Type Theory: Subset Theory." In Giovanni Sambin and Jan Smith, editors, "Twenty-Five Years of Constructive Type Theory," volume 36 of *Oxford Logic Guides*, pp. 221–244. Oxford: Oxford University Press

Emily Scida, 2004. *The Inflected Infinitive in Romance Languages*. London: Taylor & Francis

John R. Searle, 1969. *Speech Acts: An Essay in the Philosophy of Language*. Cambridge: Cambridge University Press

John R. Searle, 1983. *Intentionality: An essay in the philosophy of mind*. Cambridge: Cambridge University Press

John R. Searle, 1995. *The Construction of Social Reality*. London: The Penguin Press

John R. Searle, 2002. *Consciousness and language*. Cambridge: Cambridge University Press

John R. Searle, 2004. *Mind: A Brief Introduction*. Oxford: Oxford University Press

John R. Searle, 2010. *Making the Social World: The Structure of Human Civilization*. Oxford: Oxford University Press

Robin Beth Shamuel, editor, 2009. *The Volume of the 4th Syriac Language Conference*. Duhok (Iraq): Beth Mardootha

Mark Steedman and Jason Baldridge, 2007. "Combinatory Categorial Grammar." In R. Borsley and K. Borjars, editors, "Non-Transformational Syntax," London: Blackwell. Draft 5.0. To appear

Gabriele Stein, 2007. *A Dictionary of English Affixes – Their function and meaning*. München: Lincom Europa

John R. Taylor, 2002. *Cognitive Grammar*. Oxford: Oxford University Press

Lucien Tesnière, 1959. *Éleménts de syntaxe structurale*. 8 rue de la Sorbonne, Paris: Editions Klincksieck

Paul Thagard, 2005. "Being Interdisciplinary: Trading Zones in Cognitive Science." In Derry et al. (2005)

Michel Tomasello, 2003. *Constructing a language: A Usage-Based Theory of Language Acquisition*. Cambridge, MA. and London: Harvard University Press

Maddalena Toscano and Simon Sewangi, 2005. "Discovering Usage Patterns for the Swahili *amba-* Relative Forms c. 16, 17, 18: Using Corpus Data to Support Autonomous Learning of Kiswahili by Italian Speakers." *Nordic Journal of African Studies* 14 (3): 274–317

Mauro Tosco, 2006. "Towards a geometry of adpositional systems: A preliminary investigation of Gawwada." In Borbone et al. (2006), pp. 695–702

Anne Sjerp Troelstra, 1977. "Aspects of Constructive Mathematics." In Jon Barwise, editor, "Handbook of Mathematical Logic," volume 90 of *Studies in Logic and the Foundations of Mathematics*, chapter D.6, pp. 973–1052. Amsterdam: Elseviers

Anne Sjerp Troelstra and Helmut Schwichtenberg, 2000. *Basic Proof Theory*, volume 43 of *Cambridge Tracts in Theoretical Computer Science*. Cambridge: Cambridge University Press, 2nd edition

Edda Weigand, 2011. *Dialogue: The Mixed Game*. Amsterdam: John Benjamins

Joseph Weizenbaum, 1966. "ELIZA, a computer program for the study of natural language communication between man and machine." *Commun. ACM* 9: 36–45. ISSN 0001-0782. doi:http://doi.acm.org/10.1145/365153.365168. URL `http://doi.acm.org/10.1145/365153.365168`

Benjamin Lee Whorf, 1945. "Grammatical Categories." *Language* 21 (1): 1–11. URL `http://www.jstor.org/stable/410199`

Anna Wierzbicka, 1988. *The Semantics of Grammar*. Amsterdam: John Benjamins

Seikô Yamasaki, 2000. *Enkonduko en la japanan*. Chapecó, Brasil: Fonto

Index

Entries are ordered alphabetically; Greek letters are ordered according to their pronunciation, e.g., λ is listed under L.
Some entries are marked with ► : this sign indicates to see another entry, listed below. Some entries refers to other, related entries: these references are introduced by *also* and the entry they refer to.
Finally, entries are referred to as ► *A* ▹ *B* ▹ *C*, meaning that the referred entry is *C* which is listed under the entry *A* and the sub-entry *B*.

D

G

H

I

N

O

P

Q

R

S

T